KU-755-082

# PRINCIPLES OF HORTICULTURE

Second Edition

C.R. Adams, BSc (Agric) Hons, MIHort, Dip Applied Ed
K.M. Bamford, BSc Hons (Agric Sci), Cert Ed
M.P. Early, MSc, BSc Hons, DTA, Cert Ed

BUTTERWORTH
HEINEMANN

Butterworth-Heinemann
Linacre House, Jordan Hill, Oxford OX2 8DP
A division of Reed Educational and Professional Publishing Ltd

ℛ A member of the Reed Elsevier plc group

OXFORD    BOSTON    JOHANNESBURG
MELBOURNE    NEW DELHI    SINGAPORE

First published 1984
Reprinted 1985, 1987, 1988, 1990, 1991, 1992
Second edition 1993
Reprinted 1994, 1995, 1996, 1997

© C. R. Adams, K. M. Bamford and M. P. Early 1984, 1993

All rights reserved. No part of this publication
may be reproduced in any material form (including
photocopying or storing in any medium by electronic
means and whether or not transiently or incidentally
to some other use of this publication) without the
written permission of the copyright holder except in
accordance with the provisions of the Copyright,
Designs and Patents Act 1988 or under the terms of a
licence issued by the Copyright Licensing Agency Ltd,
90 Tottenham Court Road, London, England W1P 9HE.
Application for the copyright holder's written permission
to reproduce any part of this publication should be addressed
to the publishers

| LLYSFASI COLLEGE | |
|---|---|
| C000180365 | |
| Morley Books | 10.9.97 |
| 635 | £14.99 |
| 5561 | 2158 |

**British Library Cataloguing in Publication Data**
Adams, C. R.
    Principles of Horticulture. - 2 Rev. ed
    I. Title  II. Bamford, K. M.  III. Early, M. P.
    635

ISBN 0 7506 1722 5

Printed and bound in Great Britain by The Bath Press, Bath

# Contents

# Preface

By studying the principles of horticulture, one is able to learn how and why plants grow and develop. In this way, horticulturists are better able to understand the responses of the plant to various conditions, and therefore to perform their function more efficiently. They are able to *manipulate* the plant so that they achieve their own particular requirements of maximum yield and/or quality at the correct time.

The text therefore introduces **the plant** in its own right, and explains how a correct naming method is vital for distinguishing one plant from another. The internal structure of the plant is studied in relation to the functions performed in order that we can understand why the plant takes its particular form.

The environment of a plant contains many variable factors, all of which have their effects, and some of which can dramatically modify growth and development. It is therefore important to distinguish the effects of these factors in order to have precise control of growth. The environment which surrounds the parts of the plant above the ground includes factors such as light, day-length, temperature, carbon dioxide, oxygen, and all of these must ideally be provided in the correct proportions to achieve the type of growth and development required. The growing medium is the means of providing nutrients, water, air, and usually anchorage for the plants.

In the wild, a plant will interact with other plants, often of different species, and other organisms to create a balanced community. Ecology is the study of this balance. In growing plants for our own ends we have created a new type of community which creates problems — problems of competition for the environmental factors between one plant and another of the same species, between the crop plant and a weed, or between the plant and a pest or disease organism. These latter two competitive aspects create the need for **crop protection**.

It is only by identification of these competitive organisms (weeds, pests and diseases) that the horticulturist may select the correct method of control. With the larger pests there is little problem of recognition, but the smaller insects, mites, nematodes, fungi and bacteria are invisible to the naked eye and, in this situation, the grower must rely on the **symptoms** produced (type of damage). For this reason, the pests are covered under major headings of the organism, whereas the diseases are described under symptoms. Symptoms (other than those caused by an organism) such as frost damage, herbicide damage and mineral deficiencies may be confused with pest or disease damage, and reference is made in the text to this problem. Weeds are broadly identified as perennial or annual problems. References at the end of each chapter encourage students to expand their knowledge of symptoms.

In an understanding of crop protection, the **structure** and **life cycle** of the organism must be emphasized in order that specific measures, e.g. chemical control, may be used at the correct time and place to avoid complications such as phytotoxicity, resistant pest production or death of beneficial organisms. For this reason, each weed, pest and disease is described in such a way that **control** measures follow logically from an understanding of its biology. More detailed explanations of **specific** types of control, such as biological control, are contained in a separate chapter where concepts such as economic damage are discussed.

This book is not intended to be a reference source of weeds, pests and diseases; its aim is to show the *range* of these organisms in horticulture. References are given to texts which cover symptoms and life cycle stages of a wider range of organisms. Latin names of species are included in order that confusion about the varied common names may be avoided.

**Growing media** include soils and soil substitutes such as composts, aggregate culture and nutrient film technique. Usually the plant's water and mineral requirements are taken up from the growing medium by roots. Active roots need a supply of oxygen, and therefore the root environment must be managed to include aeration as well as to supply water and minerals. The growing medium must also provide anchorage and stability, to avoid soils that 'blow', trees that uproot in shallow soils or tall pot plants that topple in lightweight composts.

The components of the soil are described to enable satisfactory root environments to be produced and maintained where practicable. Soil conditions are modified by cultivations, irrigation, drainage and liming, while fertilizers are used to adjust the nutrient status to achieve the type of growth required.

The use of soil substitutes, and the management of plants grown in pots, troughs, peat bags and other containers where there is a restricted rooting zone, are also discussed in the final chapter.

Throughout this second edition of the book there has been a determination by the authors to provide an integrated approach across the whole span of the text, and to provide practical examples related to the growing of plants in all aspects of horticulture. More emphasis has been placed on the diversity of plant use in the sphere of horticulture. A new first chapter has been included to show horticulture in context. Some topics, such as conservation and organic growing, which have gained more prominence, have been enlarged upon and the opportunity has been taken to update in line with developments in the industry.

The indexing and key word cross-referencing has been improved to help the reader integrate the subject areas and to pursue related topics without laborious searching. It is hoped that this will enable readers to start their studies at almost any point, although it is recommended that Chapter 1 is read first. Practical exercises have been included to suggest ways to study further the concepts introduced and relate them to the practice of growing plants.

The text of this new edition has been amended to support specifically students studying for the National Certificate in Horticulture, National Diplomas in Horticulture and the RHS General Examination in Horticulture. The knowledge and understanding acquired from this textbook will be of value in preparation for Vocation Qualification (NVQ and SVQ) assessments up to Level III. Furthermore, the content of the text will provide an excellent base for those studying the General National Vocational Qualification (GNVQ) in Land Based Industries. It is also intended to be a comprehensive source of information for the keen gardener, especially for those tackling City and Guilds Amateur Gardening modules.

# Acknowledgements

We are indebted to the following people:

Marjorie Adams and Ann Taylor have shown tremendous patience, skill, and fortitude in interpreting our script for typing.

Gill Parks has developed and printed many of the photographs, and we are also grateful to Jon Parkin for the microscope photographs of plant structure.

Drs J. Bridge, D. Govier and S. Dowbiggin read the early drafts and provided valuable suggestions.

Marion Brown very kindly read the proofs.

Thanks are also due to the following individuals, firms, and organizations who provided photographs and tables:

Agricultural Lime Producers' Association.
Dr C.C. Doncaster, Rothamsted Experimental Station.
Dr P.R. Ellis, National Vegetable Research Station.
Mrs P. Evans, Rothamsted Experimental Station.
Dr D. Govier, Rothamsted, Experimental Station.

Dr M. Hollings, Glasshouse Crops Research Institute.
Dr M.S. Ledieu, Glasshouse Crops Research Institute.
Ministry of Agriculture, Fisheries and Food.
Muntons Microplants Ltd, Stowmarket, Suffolk.
Shell Chemicals.
Soil Survey of England and Wales.
Dr E. Thomas, Rothamsted Experimental Station.

Two figures illustrating weed biology and chemical weed control are reproduced after modification, with permission of Drs H.A. Roberts, R.J. Chancellor and J.M. Thurston and those illustrating the carbon and nitrogen cycles are adapted from diagrams devised by Dr E.G. Coker.

Thanks for help in revision for a second edition are due to:

Christine Atkinson, Louth Library.
John Salter, Elmwood College, Cupar, Fife.
J.B. Truscott, Lincolnshire College of Agriculture and Horticulture.

# 1

# Putting Horticulture in Context

*The many facets of horticulture have much in common, each of being concerned with the growing of plants. Despite the wide range of the industry, embracing as it does activities from the preparation of a cricket square to the production of uniformly-sized cucumbers, there are common principles which guide the successful management of the plants involved. This chapter puts the industry, the plant, and ecology into perspective and looks forward to the more detailed explanations of the following chapters.*

Horticulture may be described as the practice of growing and manipulating plants in a relatively intensive manner. This contrasts with agriculture which, in most Western European countries, relies on a high level of machinery use over an extensive area of land, consequently involving few people in the production process. However, the boundary between the two is far from clear, especially when considering large-scale *vegetable production*. There is also a fundamental difference between productive horticulture, whether producing plants themselves or plant products, and service horticulture, i.e. the development and upkeep of gardens and landscape for their amenity, cultural and recreational values. Where the tending of plants for leisure moves from being horticulture to countryside management is another moot point.

In contrast, the change associated with replacing plants with alternative materials, as in the creation of artificial playing surfaces, tests what is meant by horticulture in a quite different way.

This book concerns itself with the principles underlying the growing of plants in the following sectors of horticulture:

- *Turf culture*, which includes decorative lawns and sports surfaces for football, cricket, golf etc.
- *Garden construction and maintenance*, which involves the skills of landscaping together with the development of planted areas. Closely associated with this aspect is the maintenance of trees and woodlands (arboriculture), specialist features within the garden such as walls and patios (hard landscaping) and the use of water (aquatics). *Interior landscaping* is the provision of semi-permanent plant arrangements inside conservatories, offices and many public buildings and involves the skills of careful plant selection and maintenance.
- *Greenhouse production*, which enables plant material to be supplied outside its normal availability, e.g. chrysanthemums all the year round, tomatoes to a high specification over an extended season, and cucumbers from an area where the climate is not otherwise suitable. Plant propagation, providing seedlings and cuttings, serves outdoor growing as well as the greenhouse industry. Protected culture,

mainly using low or walk-in polythene tunnels, is increasingly important in the production of vegetables, salads and flowers.

- *Nursery stock* is concerned with the production of soil-grown or container-grown shrubs and trees. Young stock of fruit may also be established by this sector for sale to the fruit growers: *soft fruit* (strawberries etc.), *cane fruit* (raspberries etc.) and *top fruit* (apples, pears etc.).

The *private garden* may reflect many aspects of the areas of horticulture described and often embraces both the decorative and productive aspects.

## THE PLANT

There is a feature common to all these aspects of horticulture: the gardener or grower needs to know all the factors which may increase or decrease the plant's growth and development. The main aim of this book is to provide an understanding of how these factors contribute to the

ideal performance of the plant in particular circumstances. In most cases this will mean optimum growth, as in the case of a salad crop such as lettuce. But the aim may equally be restricted growth, as in the production of dwarf chrysanthemum pot plants or in the case of a lawn that requires frequent cutting. The main factors to be considered are summarized in Figure 1.1, which shows where in this book each is discussed.

It must be stressed that the incorrect functioning of any one factor may result in undesired plant performance. It should also be understood that factors such as the soil conditions, which affect the underground parts of the plant, are just as important as those such as light, which affect the aerial parts.

A single plant growing in isolation with no competition is as unusual in horticulture as it is in nature. However, specimen plants such as leeks, marrows and potatoes, lovingly reared by enthusiasts looking for prizes in local shows, grow to enormous sizes when freed from competition. In landscaping, specimen plants are placed away

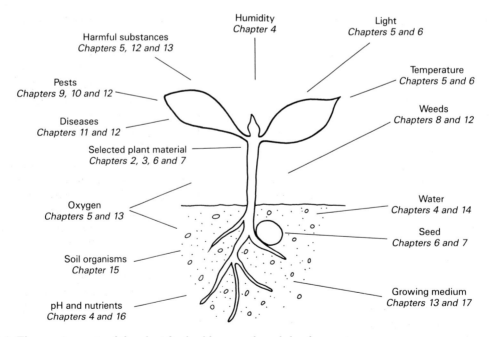

**Figure 1.1** *The requirements of the plant for healthy growth and development.*

from the influence of others so that they not only stand out and act as a point of focus, but also can attain perfection of form. A pot plant such as a fuchsia is isolated in its container, but the influence of other plants, and the consequent effect on its growth, depend on spacing. Generally, plants are to be found in groups, or communities.

## PLANT COMMUNITIES

Neighbouring plants can have a significant effect on each other since there is competition for factors such as root space, nutrient supply and light. As in natural plant communities, some of the effects can be beneficial whilst others are detrimental to the achievement of horticultural objectives.

### Single species communities

When a community is made up of one species it is referred to as a monoculture. On a football field there may be only ryegrass (*Lolium* species) with all plants closely spaced just a few millimetres apart. Each plant species, whether growing in the wild or in the garden, may be considered in terms of its own characteristic spacing distance (or plant density).

In a decorative border, the bedding alyssum will be spaced at 15 cm intervals whereas a *Pelargonium* may require 45 cm between plants. For decorative effect, the larger plants are normally placed towards the back of the border and at a wider spacing.

In a field of potatoes, the plant spacing will be closer within the row (40 cm) than between the rows (70 cm) so that suitable soil ridges can be produced to encourage tuber production, and machinery can pass unhindered along the row.

In nursery stock production, small trees are often planted in a square formations with a spacing ideal for the plant species, e.g. the conifer *Chamaecyparis* at 1.5 metres. The recent trend in producing commercial top fruit, e.g. apples, is towards small trees (using dwarf rootstocks) in order to produce manageable plants with easily harvested fruit. This has resulted in spacing reduced from 6 metres to 4 metres.

A correct plant spacing distance is that most likely to provide the requirements shown in Figure 1.1 at their optimum level. Too much competition for soil space by the roots of adjacent plants, or for light by their leaves, would quickly lead to reduced growth, and therefore reduced quality. Three ways of overcoming this problem may be seen in the horticulturist's activities of transplanting seedlings from trays into pots: increasing the spacing of pot plants in greenhouses; and hoeing out a proportion of young vegetable seedlings from a densely sown row. An interesting horticultural practice which reduces root competition is the deep-bed system, in which a one metre depth of well-structured and fertilized soil enables deep root penetration.

Whilst spacing is a vital aspect of plant growth, it should be realized that the grower may need to adjust the physical environment in one of many other specific ways in order to favour a chosen plant species. This may involve the selection of the correct light intensity; a rose, for example, whether in the garden, greenhouse or conservatory, will respond best to high light levels, while a fern will grow better in low light.

Another factor may be the artificial alteration of day length, as in the use of 'black-outs' in the commercial production of chrysanthemums to induce flowering. Correct soil acidity (pH) is a vital aspect of good growing: heathers prefer high acidity, whilst saxifrages grow more actively in non-acid (alkaline) soils. Soil texture e.g. on golf greens may need to be adjusted to a loamy sand type at the time of green preparation in order to reduce compaction and maintain drainage.

Each crop species has particular requirements, and it requires the skill of the horticulturist to bring all these together. In greenhouse production, sophisticated control equipment may monitor air and root-medium conditions every few minutes, in order to provide the ideal day and night requirements.

**Competition between species**

The subject of 'ecology' deals with the inter-relationship of plant (and animal) species and their environment. Opportunist species such as annual weeds, e.g. chickweed, grow rapidly in ground where a temporary space has been provided (e.g. by ploughing, herbicide treatment, high fertilizer application or overgrazing). These species present ongoing and often serious problems in many horticultural situations.

Combinations of crop species growing together are a common occurrence in horticulture. Whilst much of Western Europe grows crops in mono-cultures, in many other parts of the world **inter-planting** of several different species of crops, e.g. beans together with potatoes and cabbages, results in reduced attack by pests and disease. This is because an organism specifically damaging to one crop, e.g. blight on potatoes, will now have to travel further from potato plant to potato plant. A further advantage lies in the nitrogen continu-ously contributed to the soil by legume crops such as beans. A disadvantage of interplanting can be seen in its unsuitability for mechanical harvesting of crops.

In turf culture the choice of several species of grass, e.g. fescue, meadow grass and bent in a lawn seed mixture, represents an example of 'interplanting', the combination endowing the lawn with improved resistance to wear, and over-all disease resistance.

An increasingly popular practice in private gardening and organic production is the use of **companion planting**. Here one species, e.g. leeks, is grown next to another, such as carrots, specifi-cally because the two species enable each other to grow larger. The explanation behind this phenomenon may lie in the mutual provision of nutrients in the soil, the production of beneficial gaseous growth substances, or the release of gases which protect the companion species from pests. Lists of compatible species are now available to the horticulturist.

The term '**competitive**' is used in ecology to describe those species that, under natural con-ditions, out-compete the opportunist species over a long period. If horticultural land is left undisturbed for a period of five years, annual weeds give way to species such as perennial thistle, bramble and (eventually) trees such as oak. This process of change from one group of species to another is termed **succession**, and the final com-position of species ('climax species') will vary with type of soil, climate and wetness of the area (habitat). An interesting feature of succession is that, as time passes, the habitat acquires a greater diversity of species including the rotting organisms such as the fungi which break down ageing and fallen trees. Countryside management requires a knowledge of the principles of plant competition, particularly those relating to the provision of species diversity, varied habitats and selective removal of woodland tree species.

## CONSERVATION

The ecological aspects of horticulture have been highlighted in recent years by the **conservation** movement which promotes the growing of crops and maintaining of wildlife areas in such a way that the natural diversity of wild species of both plants and animals is maintained alongside crop production, with a minimum input of fertilizers and pesticides. Major public concern has focused on the effects of intensive production (mono-culture) and the indiscriminate use by horticul-turists and farmers of pesticides and quick-release fertilizers.

An example of wild-life conservation is the conversion of an area of regularly mown and 'weedkilled' grass into a wild flower meadow, providing an attractive display during several months of the year. The conversion of productive land into wild flower meadow requires lowered soil fertility (in order to favour wild species estab-lishment and competition), a choice of grass seed species with low opportunistic properties, and a mixture of selected wild flower seed. The main-tenance of the wild flower meadow may involve harvesting the area in July, having allowed time for natural flower seed dispersal. After a few years, butterflies and other insects become established as part of the wild flower habitat.

The horticulturist has two notable aspects of conservation to consider. Firstly, there must be no wilful abuse of the environment in horticultural practice. Nitrogen fertilizer used to excess has been shown, especially in porous soil areas, to be washed into streams, since the soil has little ability to hold on to this nutrient. The presence of nitrogen in watercourses encourages abnormal multiplication of microscopic plants (algae) which, on decaying, remove oxygen sources needed by other stream life, particularly fish. Another aspect of good practice increasingly expected of horticulturists is the intelligent use of pesticides. This involves a selection of those materials least toxic to man and beneficial animals, and particularly excludes those materials that increases in concentration along a food chain. The irresponsible action of allowing pesticide spray to drift onto adjacent crops, woodland or rivers has decreased considerably in recent years. This has in part been due to the Food and Environment Protection Act 1985 which has helped raised the horticulturist's awareness of conservation.

The second aspect of conservation is the deliberate selection of trees, features and areas which promote a wider range of species in a controlled manner. A golf course manager may set aside special areas with wild flowers adjacent to the fairway, preserve wet areas and plant native trees. Planting bush species such as hawthorn, field maple and spindle together in a hedgerow provides variety and supports a mixed population of insects for cultural control of pests. Tit and bat boxes in private gardens, an increasingly common sight, provide attractive homes for species which help in pest control. Continuous hedgerows will provide safe passage for mammals. Strips of grassland maintained around the edges of fields form a habitat for small mammal species as food for predatory birds such as owls. Gardeners can select plants for the deliberate encouragement of desirable species (nettles and Buddleia for butterflies; Rugosa roses and Cotoneaster for winter feeding of seed-eating birds; poached-egg plants for hoverflies).

It is emphasized that **the development and maintenance of conservation areas requires continuous management and consistent effort to maintain the desired balance of species and required appearance of the area**. As with gardens and apple orchards, any lapse in attention will result in invasion by unwanted weeds and trees.

In a wider sense the conservation movement is addressing itself to loss of certain habitats and the consequent disappearance of endangered species such as orchids from their native areas. Horticulturists are involved indirectly because some of the peat used in growing media is taken from lowland bogs much valued for their rich variety of vegetation. Conservationists also draw attention to the thoughtless neglect and eradication of wild-ancestor strains of present-day crops, a gene-bank on which future plant breeders can draw for further improvement of plant species.

## ORGANIC GROWING

The **organic movement** broadly believes that crops and ornamental plants should be produced with as little disturbance as possible to the balance of microscopic and larger organisms present in the soil, and also in the above-soil zone. This stance can be seen as closely allied to the conservation position, but with the difference that the emphasis here is on the balance of micro-organisms. Organic growers maintain soil fertility by the incorporation of animal manures, or green manure crops such as grass–clover leys. The claim is made that crops receive a steady, balanced release of nutrients through their roots; in a soil where earthworm activity recycles organic matter deep down, the resulting deep root penetration allows an effective uptake of water and nutrient reserves.

The use of most pesticides and quick-release fertilizers is said to be the main cause of species imbalance, and formal approval for licensed organic production may require soil to have been free from these two groups of chemicals for at least two years. Control of pests and diseases is achieved by a combination of resistant cultivars and 'safe' pesticides derived from plant extracts, by careful rotation of plant species, and by the use of naturally occurring predators and parasites. Weeds are controlled by mechanical and heat-producing weed-controlling equipment, and by the use of mulches. The balanced nutrition of the crop is said to induce greater resistance to pests and diseases, and the taste of organically grown food is claimed to be superior to that of conventionally grown produce.

The organic production of food and non-edible crops at present represents about 2 per cent of the European market. The European Community Regulations (1991) on the 'organic production of agricultural products' specify the substances that may be used as 'plant-protection products, detergents, fertilizers, or soil conditioners' (see pages 121 and 166). 'Conventional horticulture' is, thus, still by far the major method of production and

this is reflected in this book. However, it should be realized that much of the subsistence cropping and animal production in the Third World could be considered 'organic'.

## FURTHER READING

Allaby, M. *Concise Oxford Dictionary of Botany* (Oxford University Press, 1992).

Baines, C., and Smart, J. *A Guide to Habitat Creation* (Packard Publishing, 1991).

Blake, F. *Organic Farming and Growing* (Crowood Press, 1987).

Carr, S., and Bell, M. *Practical Conservation* (Open University/Nature Conservation Council, 1991).

Dowdeswell, W.H. *Ecology Principles and Practice* (Heinemann Educational Books, 1984).

King, T.J. *Ecology* (Thomas Nelson, 1980).

Lampkin, N. *Organic Farming* (Farming Press, 1990).

Lisansky, S.G., Robinson, A.P., and Coombs, J. *Green Growers Guide* (CPL Scientific Ltd, 1991).

# 2

# Plant Classification and Naming

*The pressures of evolution have produced widely varying plant types, found in differing habitats, and distinguished by characteristic structures and modes of life. An orderly system of classification enables the horticulturist, plant scientist and naturalist to identify any plant from any country. A detailed procedure for naming organisms is therefore essential.*

## PRINCIPLES OF CLASSIFICATION

The organisms constituting the plant kingdom are distinguishable from animals by features of nutrition and sedentary growth habits. Two other plant characters are the cellulose cell wall and polyploidy (*see* Chapter 7). **Any classification system involves the grouping of organisms or objects using characteristics common to members within the group**. Various systems have been devised throughout history, but a seventeenth-century Swedish botanist, **Linnaeus**, laid the basis for much subsequent work in the classification of plants, animals, and also minerals. In order to produce a universally acceptable system, the International Code of Botanical Nomenclature has been formulated, which includes both non-cultivated plants and details specific to cultivated plants.

The **divisions** of the plant kingdom are the main groupings of organisms according to their place in evolutionary history. Simple single-celled organisms from aquatic environments evolved to more complex descendants, multicellular plants with diverse structures, which were able to survive in a terrestrial habitat and develop sophisticated reproduction mechanisms. Most of the evolutionary stages in this development are represented in the divisions of the plant kingdom existing today. Further grouping, within each division (the names ending in -phyta), include **class** (ending -ae or -mycetes in fungi), **order** (ending -ales), **family** (ending -aceae), **genus** and **species**. **Species is the basic unit of classification, and is defined as 'a group of individuals with the greatest mutual resemblance, which are able to breed amongst themselves'**. A number of species with basic similarities constitute a genus (plural genera), a number of genera a family, a number of families an order.

### Divisions of the plant kingdom

*Viruses* are often called living organisms, but are not normally termed plants or animals as their characteristics are unusual. They are visible only under an electron microscope, and do not have a cellular structure, but consist of nucleic acid and protein. Viruses survive by invading the cells of other organisms, modifying their behaviour and causing disease in some cases.

*Bacteria* are single-celled organisms sometimes arranged in chains or groups. They are not usually included in plant classification, but have great importance to horticulture by their beneficial activities in the soil, and as causitive organisms of plant diseases.

*Algae*, comprising some 18 000 species, are true plants, since they use chlorophyll to photosynthesize (*see* Chapter 5). The division Chlorophyta (green algae) contains single-celled organisms which require water for reproduction and can present problems when blocking irrigation lines and clogging water tanks. Marine algal species in Phaeophyta (brown algae) and Rhodophyta (red algae) are multicellular, and have leaf-like structures. They include the seaweeds, which accumulate mineral nutrients, and are therefore a useful source of compound fertilizer as a liquid feed.

*Fungi*, or Mycophyta, are considered to be a division of the plant kingdom with many characteristics of plants, but they do not photosynthesize. They must, therefore, obtain their food directly from other living organisms, possibly causing disease (*see* Chapter 11), or from dead organic matter, so contributing to its breakdown in the soil.

*Lichens* form the division Lichenes, but their classification is complex since each lichen consists of both fungal and algal parts. Both organisms are mutually beneficial or symbiotic. The significance of lichens to horticulture is not great. Of the 15 000 species, one species is considered a food delicacy in Japan. However, lichens growing on tree bark or walls are very sensitive to atmospheric pollution, particularly to the sulphur dioxide content of the air. Different lichen species can withstand varying levels of sulphur dioxide, and a survey of lichen species can be used to indicate levels of atmospheric pollution in a particular area. Lichens are also used as a natural dye, and can form an important part of the diet of some deer.

*Mosses and liverworts*. The 25 000 species are included in the division Bryophyta, and have distinctive vegetative and sexual reproductive structures, the latter producing spores which require damp conditions for survival. The low spreading carpets of vegetation present a weed problem on the surface of compost in container-grown plants, on capillary benches, and around glazing bars on greenhouse roofs.

*Ferns and horsetails*, in the division Pteridophyta, represent 15 000 species which have identifiable leaf, stem and root organs, but produce spores from the sexual reproduction process. Many species of ferns, e.g. maidenhair fern (*Adiantum cuniatum*), and some tropical horsetails, are grown for decorative purposes, but the common horsetail (*Equisetum arvense*), and bracken (*Pteris aquilina*) which spread by underground rhizomes, are difficult weeds to control.

*Seed-producing plants* (Spermatophyta) contain the most highly evolved and structurally complex plants. There are species adapted to most habitats and extremes of environment. Sexual reproduction produces a seed which is a small, embryo plant contained within a protective layer.

The class **Gymnospermae** has approximately 700 surviving species which produce 'naked' seeds, usually in cones, the female organ. This class shows some primitive features, and often displays structural adaptations to reduce water loss (*see* Figure 4.2). The order Ginkgoales is represented by a single surviving species, the maidenhair tree (*Ginkgo biloba*), which has an unusual slit-leaf shape, and yellow colour in autumn. The most important order to horticulture, the conifers (Coniferales), provides many families with horticulturally interesting plant habits, and foliage shape and colours. The Cupressaceae, for example, include fast-growing species which can be used as windbreaks, and small, slow-growing types very useful for rock gardens. Taxaceae, a highly poisonous family, contains the common yew (*Taxus baccata*) used in ornamental hedges and mazes.

The second class of plants producing seeds, this time protected by fruits, the **Angiospermae**, have flower structures as the means of sexual reproduction, and this characteristic structure is used as the basis of their classification. There are estimated to be some 25 000 species, occupying a very wide range of habitats. Many Angiosperm families are important to horticulture, both as crop plants and weeds. The sub-class **Monocotyledonae** contains some horticultural families, e.g. Liliaceae, with plants such as tulips and onions, Amaryllidaceae or daffodil family, Iridaceae or Iris family, and Graminae with all the grass species. The sub-class **Dicotyledonae** has many more families significant to horticulture, including Compositae, the members of which have a characteristic flower head with many small florets making up the composite, regular (or actinomorphic) structure, e.g. chrysanthemum, groundsel. Cruciferae, characterized by its four-petal flower, contains the Brassica genus with a number of important crop plants such as cabbage, Brussels sprouts and wallflower (*Cheiranthus cheiri*). Most of the Brassicas have a **biennial growth habit, when their life cycles span over two growing seasons**, producing vegetative growth in the first season, and flowers in the second, usually in response to a cold stimulus (*see* vernalization). A number of weed species are found in the Crucifer family, including shepherd's purse (*Capsella bursa-pastoris*), which is an **annual**, **and completes its life cycle in one growing season**. Many important genera, e.g. apples (*Malus*), pear (*Pyrus*), rose (*Rosa*), are found within the Rosaceae family, which generally produces succulent fruit from a flower with five petals and often many male and female organs. Many species within this family display a **perennial growth habit, having life cycles which continue over a number of growing seasons**. Leguminosae (Fabaceae) (pea and bean family) have five-petalled asymmetric or zygomorphic (having only one plane of symmetry) flowers, which develop into long pods (legumes) containing starchy seeds. Modified leaflets (tendrils) enable an erect climbing habit in some species. The characteristic upturned umbrella-shaped flower head or **umbel** is found in the Umbelliferae (carrot family), and bears small white five-petalled flowers which are wind-pollinated.

Many other families contain horticulturally important species, e.g. Solanaceae includes potatoes and tomatoes; Cucurbitaceae, cucumbers and marrows; and Labiatae, many species used as culinary herbs. Cactaceae and Crassulariaceae are widely used as ornamental house plants, many species having stems and leaves adapted for water storage. The Aceracea include the common sycamore and a number of ornamental *Acers*, while the Salicaceae family represents the willow species.

## NAMING OF CULTIVATED PLANTS

### The binomial system

The common names which we use for plants, e.g. potato for *Solanum tuberosum*, lettuce for *Lactuca sativa* are, of course, acceptable in English, but are not universally used. **Linnaeus** formulated a system which he claimed should identify an individual plant type, by means of the composed **genus** name, followed by the **species** name. For example, the chrysanthemum used for cut flowers is *Chrysanthemum* genus and *morifolium* species; note that the genus name begins with a capital letter, while the species has a small letter. Other examples are *Ilex aquifolium*, *Magnolia stellata*, *Ribes sanguineum*.

However, these two words may not encompass all possible variations, since a species can give rise to a number of naturally occurring **varieties** with distinctive characteristics. In addition, cultivation, selection and breeding have produced variation in species referred to as cultivated varieties or **cultivars**. The two terms, variety and cultivar, are exactly equivalent, but the botanical variety name is referred to in Latin, beginning with a small letter, while the cultivar is given a name often relating to the plant breeder who produced it. There is no other significant difference in the use of the two terms, and therefore either is acceptable. However, the term cultivar will be used throughout this text. A cultivar name should be written in inverted commas and begin with a capital letter, after the binomial name or, when applicable, the common name. Examples include Prunus padus 'Grandiflora', tomato 'Sonatine', apple 'Bramley's seedling'. If a cultivar name has more than one acceptable alternative, they are said to be synonyms (sometimes written syn.).

### Hybrids

When **cross pollination** occurs between two plants, hybridization results, and the offspring usually bear characteristics distinct from either parent. Hybridization can occur between different cultivars within a species, sometimes resulting in a new and distinctive cultivar (*see* Chapter 7), or between two species, resulting in an **interspecific hybrid**, e.g. *Prunus* × *yedoensis* and *Erica* × *darleyensis*. A much rarer hybridization between two different genera results in an **intergeneric hybrid**, e.g. × *Cupressocyparis leylandii* and × *Fatshedera lizei*. The names of the resulting hybrid types include elements from the names of the parents, connected or preceded by a multiplication sign (×). A chimaera, consisting of tissue from two distinct parents, is indicated by a 'plus' sign, e.g. + *Laburnocystisus adamii*, the result of a graft.

## PRACTICAL EXERCISES

1   Observe the range of **lichen** species to be found growing on trees and walls, in relation to the proximity of industrial pollution.
2   Investigate the use of various classification groups in horticulture, according to your specialist interests, e.g. list the major families used in forestry, ornamental and commercial horticulture.
3   Examine the flower characteristics which have led to the classification of species into the various families.

## FURTHER READING

Doyle, W.T. *Non Seed Plants, Form and Function* (Macmillan, 1973).

Gledhill, D. *The Names of Plants*, 2nd edition (Cambridge University Press, 1989).

Heywood, V.H. *Plant Taxonomy*, 2nd edition (Edward Arnold, 1976).

Heywood, V.H., and Moore, D.M. *Current Concepts in Plant Taxonomy* (Academic Press, 1984).

*International Code of Nomenclature of Cultivated Plants* (International Bureau for Plant Taxonomy and Nomenclature, 1969).

Wight, F.G. *Plant Classification* (Edward Arnold, 1970).

# 3

# Plant Organization

*A multicellular organism such as the plant, which carries out many complex processes involved in its growth and development, requires a complex organization to carry out its functions. To be efficient, the plant's structural unit must be sub-divided so that each major function is carried out by a particular area in the plant, i.e. an* **organ**. *The individual units of the plant, the* **cells**, *are grouped together into* **tissues** *of similar cell types, and each tissue contributes to the activities of the whole organ.*

## PLANT FORM

Most plant species at first sight appear very similar, since all four organs, the root, stem, leaf, and flower, are present in approximately the same form and have the same major functions.

In many species the functions of the root system are to take up water and minerals from the growing medium, and to anchor the plant in the growing medium. Two types of root system are produced; **a tap root is a single large root which usually maintains a direction of growth in response to gravity** (*see* geotropism), with many small lateral roots growing from it, e.g. in chrysanthemums, brassicas, dock; a fibrous root system **consists of**

**many roots growing out from the base of the stem**, as in grasses and groundsel.

The leaf, consisting of the leaf blade (**lamina**) and stalk (**petiole**), carries out photosynthesis, its shape and arrangement on the stem depending on the water and light energy supply in the species' habitat.

The stem's function is physically to support the leaves and the flowers, and to transport water, minerals and food between roots, leaves and flowers. The leaf joins the stem at the **node** and has in its angle (axil) with the stem an **axillary bud**, which may grow out to produce a lateral shoot. The distance between one node and the next is termed the **internode** (*see* Figure 3.1).

Sexual reproduction is carried out in the flower, and therefore its appearance depends principally on the agents of pollination.

Some species have adapted to specialized habitats such as deserts or forests. Their organs have consequently evolved to most efficiently survive that environment. Plants adapted to dry areas, e.g. cacti, have leaves reduced to protective spines, and stems capable of photosynthesis. Many members of the Leguminosae possess leaves modified for climbing, while other plants use stems or roots as food and water storage organs (*see* vegetative propagation. Plants found growing in coastal areas have adaptations which allow them to withstand high salt levels, e.g. salt glands as found in *Spartina spp.* and succulence in scurvy grass). **By examining the internal structure of the**

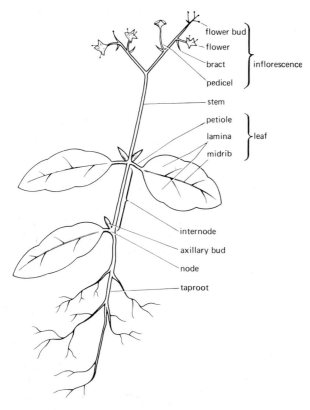

flower bud
flower
bract
pedicel
} inflorescence

stem
petiole
lamina
midrib
} leaf

internode
axillary bud
node
taproot

**Figure 3.1** *Plant stylized to show major external structures and terminology.*

**plant, the distribution of cells and tissues, and their functions, may be understood**. The stem will be used to illustrate this principle.

## STEM STRUCTURE

### Dicotyledonous stem

The internal structure of a dicotyledonous marrow stem, as viewed in cross-section, is shown in Figure 3.2. The **epidermis** is present as an outer protective layer of the stem, leaves and roots. It consists of a single layer of the individual units from which the plant is composed, the **cells**, which range from 10 to 50 micrometres in size. A small proportion of the cells are modified to allow gases to pass through an otherwise impermeable layer (*see* stomata). The two layers of cells to the inside of the epidermis are responsible for providing physical support for the young stem. All cells with a common function (together called a **tissue**) possess a similar microscopic structure (**anatomy**).

The tissues responsible for **support** in the young plant are the **collenchyma** and **sclerenchyma**. Each cell has an external **cell wall** which gives the cell its shape and rigidity. The wall's thickness will determine both a cell's rigidity and its ability to support the tissue in which it is found. Both collenchyma and sclerenchyma tissues have cells with specially thickened walls. When a cell is first formed it has a wall composed mainly of **cellulose** fibres. In collenchyma cells, the amount of cellulose is increased, but otherwise the cells remain relatively unspecialized. In sclerenchyma cells, the thickness of the wall is increased by the addition of a substance called **lignin**, which is hard and causes the living contents of the cell to disappear. These cells, which are long and tapering and interlock for additional strength, therefore consist only of cell walls.

The cortex of the stem contains a number of different tissues. Some cells have no specialized function (**parenchyma**) but are able, when required, to produce energy by respiration, photosynthesize, store food, or divide. Also contained in the cortex are **vascular bundles**, so named because they contain two vascular tissues which are responsible for transport. The first, **xylem**, contains long, wide, open-ended cells with very thick lignified walls, able to withstand the high pressures of water with dissolved minerals which they carry. The second vascular tissue, **phloem**, consists again of long, tube-like cells, and is responsible for the transport of food manufactured in the leaves and carried to the roots or stems or flowers (*see* translocation). The phloem tubes, in contrast to xylem, have fairly soft cellulose cell walls. The end-walls are only partially broken down to leave sieve-like structures at intervals along the phloem tubes. The phloem is seen on either side of the xylem in the marrow stem, but is found to the outside of the xylem in most other species. Phloem is penetrated by the stylets of feeding aphids (*see* Chapter 10).

Also contained within the vascular bundles is the **cambium** tissue, which contains actively

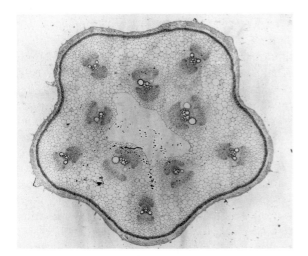

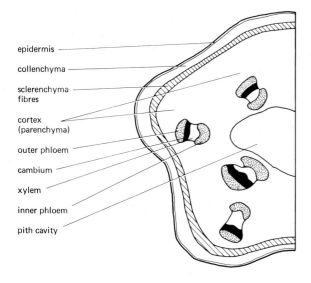

epidermis
collenchyma
sclerenchyma fibres
cortex (parenchyma)
outer phloem
cambium
xylem
inner phloem
pith cavity

**Figure 3.2** *Cross-section of young stem of marrow (Cucurbita pepo) showing areas of tissue for support (note thick cell walls) and transport (note especially large diameter xylem vessels).*

dividing cells producing more xylem and phloem tissue as the stem grows.

### The monocotyledonous stem

This has the same functions as those of a dicotyledon; therefore the cell types and tissues are similar. However, the arrangement of the tissues does differ because increase in diameter by **secondary growth** does not take place. The stem relies on extensive sclerenchyma tissue for support which, in the maize stem shown in Figure 3.3, is found as a sheath around each of the scattered vascular bundles. The absence of secondary growth in the vascular bundles makes the presence of cambium tissue unnecessary.

## STEM GROWTH

The tip of the stem or **apex**, enclosed in a terminal bud, increases in length by **cell division** in the **meristem**, cell enlargement, and eventually develops particular cell types, e.g. xylem or sclerenchyma. This apical zone consists of soft tissues which, while very active (*see* tissue culture) are also vulnerable to pest attack and therefore physical damage. If the normal growth and organization of the tissues is disturbed, a disorder results, called **fasciation**, which may resemble a number of stems fused together. In some plant families, e.g. the Graminae, the meristem remains at the base of the leaves, which are therefore protected against some herbicides, e.g. 2,4-D (*see* Chapter 12).

As the stem length increases, so width also increases to support the bigger plant and supply the greater amount of water and minerals required. The process in dicotyledons is called **secondary growth** (*see* Figure 3.4).

Additional phloem and xylem are produced on either side of the cambium tissue, which now forms a complete ring. As these tissues increase towards the centre of the stem, so the circumference of the stem must also increase. Therefore

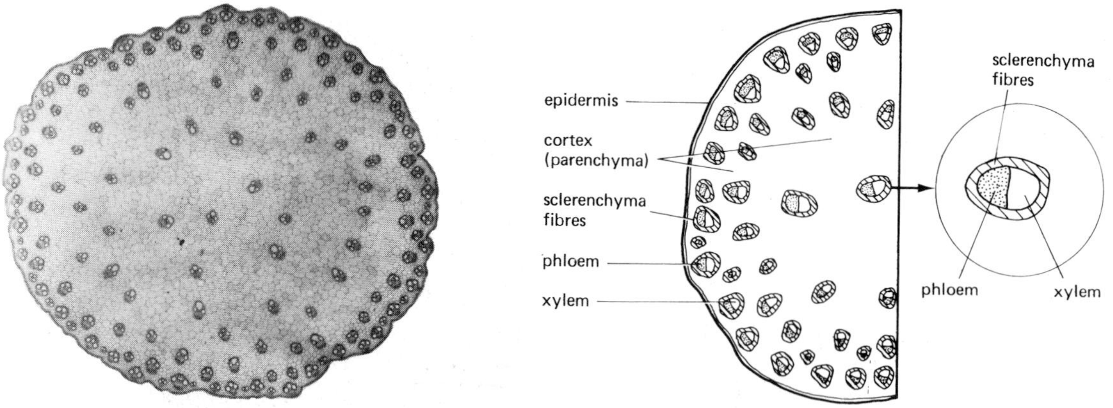

**Figure 3.3** *Cross-section of young stem of maize* (Zea mays) *showing scattered vascular bundles with associated support tissue.*

a secondary ring of cambium (cork cambium) is formed, just to the inside of the epidermis, the cells of which divide to produce a layer of corky cells on the outside of the stem. This layer will increase with the growth of the tissue inside the stem, and will prevent loss of water if cracks should occur. As more secondary growth takes place, so more phloem and xylem tissue is produced but the phloem tubes, being soft, are squashed as the more numerous and very hard xylem vessels occupy more and more of the cross-section of the stem. Eventually, the majority of the stem consists of secondary xylem which forms the **wood**. The central region of xylem sometimes becomes darkly stained with gums and resins (**heartwood**) and performs the long-term function of support for a heavy trunk or branch. The outer xylem, the **sapwood**, is still functional in transporting water and nutrients, and is often lighter in colour. The xylem tissue produced in the spring has larger diameter vessels than autumn-produced xylem, due to the greater volume of water which must be transported; a distinct ring is therefore produced where the two types of tissue meet. As these rings will be formed each season, their number can indicate the age of the branch or trunk; they are called **annual rings**. The phloem tissue is pushed against the cork layers by the increasing volume of xylem so that a woody stem

appears to have two distinct layers, the wood in the centre and the bark on the outside. If bark is removed, the phloem also will be lost, leaving the vascular cambium exposed. The stem's food transport system from leaves to the roots is thus removed and, if a trunk is completely **ringed**, the plant will die. This sort of damage may be caused by rabbits in an orchard. Reduction in growth rate of vigorous tree fruit cultivars and woody ornamental species can be achieved by 'partial ringing'.

Since secondary growth is produced by division of cells in the cambium, it is important that when grafting a **scion** (the material to be grafted) to a **stock**, the vascular cambium tissues of both components be positioned as close to each other as possible. The success of a graft depends very much on the rapid **callus** growth derived from the cambium from which new cambial cells form and subsequently from which the new xylem and phloem vessels form to complete the union. The two parts then grow as one to carry out the functions of the plant stem.

A further feature of a woody stem is the mass of lines radiating outwards from the centre, most obvious in the xylem tissues. These are **medullary rays**, consisting of parenchyma tissue linking up with small areas on the bark where the corky cells are less tightly packed together (**lenticels**). These

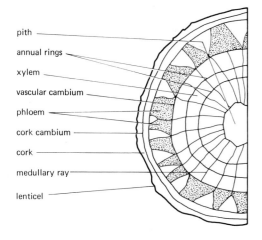

pith
annual rings
xylem
vascular cambium
phloem
cork cambium
cork
medullary ray
lenticel

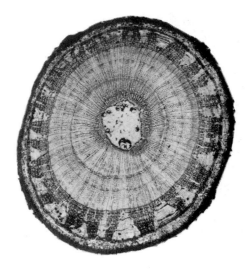

**Figure 3.4** *Cross-section of woody stem of lime* (Tilia europea) *showing large central woody area and development of bark to the outside.*

allow air to move into the stem and across the stem from cell to cell in the medullary rays. The oxygen in the air is needed for the process of **respiration**, but the openings can be a means of entry of some diseases, e.g. Fireblight. Other external features of woody stems include the **leaf scars** which mark the point of attachment of leaves fallen at the end of a growing season, and can be a point of entry of fungal spores such as apple canker.

and root cortex cells, release energy by means of respiration. Starch grains may form as a means of storage for the products of photosynthesis. **Vacuoles** act as storage areas for water and dissolved minerals, and waste products of the processes occurring in the cytoplasm. The vacuole can become a large proportion of the cell volume, and assists by pressure of water in the support of the plant's tissues. The whole living matter forms the **protoplasm** of the plant.

## CELL STRUCTURE

A simple, unspecialized cell (*see* Figure 3.5), e.g. parenchyma, consists of a cellulose cell wall, and contents (**cytoplasm**) enclosed in a **cell membrane** which is selective for the passage of materials in and out of the cell. Suspended in the jelly-like cytoplasm are small structures, each enclosed within a membrane, having specialized functions within the cell. The **nucleus** contains the **chromosomes**, which control the activities of the cell. The green **chloroplasts** present in large numbers in the **pallisade tissue** of the leaf are responsible for carrying out photosynthesis, and the **mitochondria**, found most commonly in the meristems

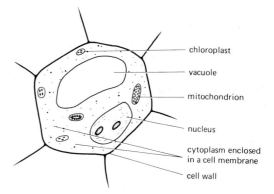

chloroplast
vacuole
mitochondrion
nucleus
cytoplasm enclosed in a cell membrane
cell wall

**Figure 3.5** *An unspecialized plant cell showing the organelles responsible for the life processes.*

## PRACTICAL EXERCISES

1  Examine the modifications of plant form found in:
   (a) the stems of cacti for water retention and photosynthesis
   (b) the leaves of peas for climbing.
2  Examine the internal structure of the stem of:
   (a) cabbage, by cutting a cross section with a sharp knife; note specially the xylem tissue
   (b) any deciduous woody species, by cutting across a twig with a sharp knife; note the wood and bark.

## FURTHER READING

Clegg, C.J., and Cox, G. *Anatomy and Activities of Plants*, Section 1 (John Murray, 1978).
Cutler, D.F. *Applied Plant Anatomy* (Longman, 1978).
Vines, A.E., and Rees, N. *Plant and Animal Biology*, Vol. 1, Ch. 12 (Longman, 1986).

# 4

## Water and Minerals in the Plant

*Water is the major constituent of any living organism, and the maintenance of a plant with an optimum water content is a very important part of plant growth and development (see Soil Water, Chapter 14). Probably more plants die from lack of water than from any other cause. Minerals are also raw materials essential to growth, and are supplied through the root system.*

### WATER

#### Functions of water

The plant consists of about 95% water, which is the main constituent of **protoplasm** or living matter. When the plant cell is full of water, or **turgid**, the pressure of water enclosed within a membrane or vacuole acts as a means of support for the cell and therefore the whole plant, so that when a plant loses more water than it is taking up, the cells collapse and the plant may wilt. Aquatic plants are supported largely by external water and have very little specialized support tissue. In order to survive, any organism must carry out complex chemical reactions which are explained, and their horticultural application described, in Chapters 5 and 6. Raw materials for chemical reactions must be transported and brought into contact with each other by a suitable medium; water is an excellent solvent. One of the most important processes in the plant is photosynthesis, and a small amount of water is used up as a raw material in this process.

### MOVEMENT OF WATER

Water moves into the plant through the roots, the stem, and into the leaves, and is lost to the atmosphere. **By the process of diffusion, molecules of a gas or liquid move from an area of high concentration to an area where there is a relatively lower concentration of the diffusing substance.** Thus, water vapour moves through the **stomata** (*see* Figure 4.2) from an area of high concentration inside the leaf into the air immediately surrounding the leaf where there is a lower relative humidity. The pathway of water movement through the plant falls into three distinct stages: water uptake, movement up the stem, and transpirational loss from the leaves.

#### 1 Water uptake

The movement of water into the roots is by a special type of diffusion called **osmosis**. Soil water enters root cells through the **cell wall** and **membrane**. The cell wall is permeable to both soil water and the dissolved inorganic minerals, but the cell membrane, although permeable to water,

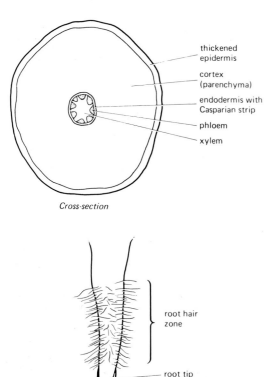

*Cross-section*

thickened
epidermis

cortex
(parenchyma)

endodermis with
Casparian strip

phloem

xylem

root hair
zone

root tip

root cap

*Long view of
root tip region*

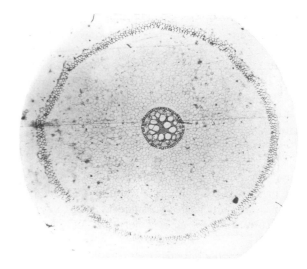

**Figure 4.1** *Cross-section of lily root showing thickened outer region, large area of cortex and central vascular region enclosed in a single celled endodermis.*

allows only the smallest molecules to pass through, somewhat like a sieve, and is therefore described as a **semi-permeable membrane**.

The minerals move into the cells by a process which requires energy to 'push' the molecules into the cell, and a greater concentration of minerals is usually maintained inside the cell compared with that in the soil water. This means that, by osmosis, water will move from the soil into the cell where there is relatively less water as there are more inorganic salts and sugars. The greater the difference in concentration of inorganic salts either side of the cell membrane, the greater the **osmotic pressure** and the faster water moves into the root cells, also affected by increased temperature. **Osmosis** can therefore be defined as the movement of water from an area of low salt concentration to

an area of relatively higher salt concentration, through a semi-permeable membrane. If there is a build-up of salts in the soil, either over a period of time or, for example, where too much fertilizer is added, water may move out of the roots by osmosis, and the cells are then described as **plasmolysed**. Cells which lose water can recover their water content if the correct conditions are rectified quickly, but this can lead to permanent wilting. Such situations can be avoided by correct dosage of fertilizer and by monitoring of conductivity levels in greenhouse soils and NFT systems (*see* Chapter 17).

**Root structure** (*see* Figure 4.1). The function of the root system is to take up water and mineral nutrients from the growing medium and to anchor the plant in that medium. Its major function involves contact with the soil water, and it must have a large surface area in order to tap as much of the growing medium's resources as possible. The root surface very near to the tip where growth occurs, protected by the **root cap**, has projections called **root hairs**, reaching numbers of 200–400 per square millimetre, which greatly increase the surface area in this region. Plants grown in nutrient solutions, e.g. NFT, produce considerably fewer root hairs. The loss of root hairs during transplanting can check plant growth considerably, and the hairs can be points of entry of diseases such as club root (*see* Chapter 11). Figure 4.1 shows that the layer with the root hairs, the **epidermis**, is comparable with the epidermis of the stem (*see* stem structure); it is a single layer of cells which has a protective as well as an absorptive function. Inside the epidermis is the parenchymatous **cortex**, the main function of which is respiration, which produces energy for growth of the root and for the absorption of mineral nutrients. The cortex can also be used for the storage of food where the root is an overwintering organ. The cortex is therefore often quite extensive, and water must move across it in order to reach the transporting tissue which is in the centre of the root. This central region, called the **stele**, is separated from the cortex by a single layer of cells, the **endodermis**, which has the function of controlling the passage of water into the stele. A waxy strip forming part of the cell wall of many of the endodermal cells, called the Casparian strip, prevents water from moving into the cell by all except the cells outside this strip, called passage cells. In this way, the volume of water passing into the stele is restricted. If such control did not occur, often more water would move into the transport system than the plant needs. In some conditions, such as in high air humidity and fluctuating temperatures, more water moves into the leaves than is being lost to the air, and the cell walls in the leaf may burst. This condition is known as **oedema**, and commonly occurs in *Pelargonium* as dark green patches becoming brown, and also weak celled plants such as lettuce, when it is known as **tip-burn**, because the margins of the leaves particularly will appear scorched. **Guttation** may occur when liquid water is forced onto the leaf surface. Water passes through the endodermis to the xylem, which transports the water and dissolved minerals up to the stem and leaves. The arrangement of the xylem tissue varies between species, but often appears in transverse section as a star with varying numbers of 'arms'. Phloem tissue is responsible for transporting carbohydrates from the leaves as a food supply for the production of energy in the cortex. A distinct area in the root inside the endodermis, the pericycle, supports cell division and produces lateral roots, which push through to the main root surface from deep within the structure.

Roots, as with stems, age and become thickened with waxy substances, and the uptake rate of water becomes restricted.

## 2 Movement of water up the stem

Although a complete scientific explanation is not available, a number of contributing forces are established. Osmotic forces, previously described, push water up the root and stem xylem by **root pressure**. A small lift is achieved in the smallest diameter xylem vessels by **capillary attraction**, which is the attraction of the water molecules for the sides of the vessel (*see* Chapter 14). The considerable mutual attraction of the water molecules enables the water columns inside the xylem to move like a solid mass and be sucked up when subjected to negative pressure from above. This movement is called **transpiration pull**.

## 3 Transpiration

**Transpiration is the loss of water vapour from the leaves of the plant.** Any plant takes up a lot of water through its roots; for example, a tree can take up about 900 litres (200 gallons) a day. Approximately 98 per cent of the water taken up moves through the plant and is lost by transpiration; only about 2 per cent is retained as part of the plant's structure, and a yet smaller amount is used up in photosynthesis. The extravagant loss through leaves is due to the unavoidably large pores in the leaf surface (**stomata**) essential for carbon dioxide diffusion (*see* Figure 4.2).

The potential for the entry of other factors, for example, **fungi**, is also due partly to the existence of these stomata. If the air surrounding the leaf becomes very humid, then the **diffusion** of water vapour will be much reduced and the rate of transpiration will decrease. Application of water to greenhouse paths during the summer, **damping down**, increases relative humidity and reduces transpiration rate. If the air surrounding the leaf is moving, humidity of air around the leaf is low, so that transpiration is maintained. Windbreaks, e.g. some woody species, reduce the risk of desiccation of crops. Ambient temperatures affect the rate at which liquid water in the leaf vaporizes (i.e. evaporates), and thus determines the transpiration rate (*see* evapotranspiration). The plant is able to control transpiration rate, as most of its surface is protected by a waxy waterproof layer or **cuticle**, and the stomata are able to close up (*see* leaf structure). The stomatal pore is bordered by two sausage-shaped guard cells, which have thick cell walls near to the pore. When the guard cells are fully turgid, the pressure of water on the thinner walls causes the cells to buckle and the pore to open. If the plant begins to lose more water, the guard cells lose their turgidity and the stomata close to prevent any further water loss. Stomata also respond to light and dark respectively by opening and closing. The mechanism of this action is not fully understood. The very quick response of the chloroplast-containing guard cells is unlikely to be due to cell turgor induced by photosynthesis. Stomata also close if carbon dioxide concentration in the air rises above optimum levels.

A close relationship exists between the daily fluctuation in the rate of transpiration and the variation in solar radiation. This is used to assess the amount of water lost from cuttings supplied with mist units (*see* misting), automatically switched by a light-sensitive cell. In artificial conditions, e.g. in a florist shop, transpiration rate can be reduced, and therefore the need for a high water supply, by providing a cool, humid and shaded environment. **Plasmolysed** leaf cells can occur if highly concentrated sprays cause water to leave the cells and result in scorching.

The evaporation of water from the cells of the leaf means that in order for the leaf to remain turgid, which is important for efficient photosynthesis, the water lost must be replaced by water in the xylem. Pressure is created in the xylem by the loss from an otherwise closed system and water moves up the petiole of the leaf and stem of the plant by suction (*see* **transpiration pull**). If the water in the xylem column is broken, for example when a stem of a flower is cut, air moves into the xylem and may restrict the further movement of water when the cut flower is placed in water. Once the column is restored, however, water enters by the cut surface at a faster rate than if the plant was intact with a root system.

**Antitranspirants** are plastic substances which, when sprayed onto the leaves, will create a temporary barrier to water loss over the whole leaf surface, including the stomata. These substances are useful to protect a plant during a critical period in its cultivation; for example, conifers can be treated while they are moved to another site.

Some species have special structural adaptations to enable them to withstand low water supplies. Figure 4.2 shows a cross-section of a *Pinus* leaf with a reduced surface area, very thick cuticle, and sunken guard cells protected below the leaf surface. In extreme cases, e.g. cacti, the leaf is reduced to a spine, and the stem takes over the

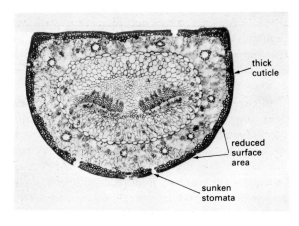

thick cuticle

reduced surface area

sunken stomata

**Figure 4.2** *Cross-section of pine* (Pinus) *leaf showing some adaptations to reduce excessive water loss.*

function of photosynthesis and is also capable of water storage.

## MINERALS

**Essential minerals** are those inorganic substances necessary for the plant to grow and develop normally, which can be conveniently divided into two groups. **The major nutrients (macronutrients) are required in relatively large quantities, and include nitrogen, phosphorus, potassium, magnesium, calcium and sulphur. The micronutrients (trace elements) including iron, boron, manganese, copper, zinc and molybdenum, are needed in relatively small quantities, usually measured in parts per million, and within a narrow concentration range to avoid deficiency or toxicity.**

**Non-essential minerals**, such as sodium and chlorine, have important functions in the plant but are not constantly required for plant growth and development.

### Functions and deficiency symptoms of minerals in the plant

Many essential minerals have very specific functions in the plant cell processes. When in short supply (**deficient**), certain characteristic symptoms are shown by the plant, but these symptoms indicate an extreme deficiency. To ensure optimal mineral supplies, growing media analysis (*see* Chapter 16) or plant tissue analysis can be used to forecast low nutrient levels, which can then be suitably increased.

**Nitrogen** is a constituent of **proteins**, nucleic acids and chlorophyll and, as such, is a major requirement for plant growth. Its compounds comprise about 50 per cent of the dry matter of protoplasm, the living substance of plant cells. **Deficiency** causes slow, spindly growth in all plants and yellowing of the leaves (chlorosis) due to lack of chlorophyll. Stems may be red or purple due to the formation of other pigments. The high mobility of nitrogen in the plant to the younger, active leaves causes the old leaves to show symptoms first.

**Molybdenum**, a trace element, assists the uptake of nitrogen, and although required in much smaller quantities, its deficiency can result in reduced plant nitrogen levels. In tomatoes and lettuce, deficiency of molybdenum can lead to chlorosis in older leaves, followed by death of cells between the veins (interveinal) and leaf margins. Tissue browning (**necrosis**) and infolding of the leaves may occur. In brassicas, the 'whiptail' leaf symptom involves a dominant midrib and loss of leaf lamina.

**Phosphorus** is important in the production of nucleic acid, and large amounts are therefore concentrated in the meristem. Organic phosphates, so vital for the plant's respiration, are particularly required in active organs such as roots and fruit, while the seed must store adequate levels for germination. **Deficiency** symptoms are not very distinctive. Poor establishment of seedlings results from a general reduction in growth of stem and root systems. Sometimes a general darkening of the leaves in dicotyledonous plants leads to brown leaf patches, while a reddish tinge is seen in monocotyledons. In cucumbers grown in deficient peat composts or NFT, characteristic stunting and development of small young leaves leads to brown spotting on older leaves.

**Potassium**. Although present in relatively large amounts in plant cells, this mineral does not have any clear function in the formation of important cell products. It acts as an osmotic regulator, for example in guard cells (*see* osmosis), and is involved in resistance to **chilling injury**, drought and disease. **Deficiency** results in brown, scorched patches on leaf tips and margins, especially on older leaves, due to the high mobility of potassium towards growing points. Leaves may develop a bronzed appearance and roll inwards and downwards.

**Magnesium** is a constituent of chlorophyll, and is involved in the activation of some **enzymes**, and in the movement of phosphorus in the plant. **Deficiency** produces characteristic interveinal chlorosis, initially in older leaves, which subsequently become reddened. Necrotic areas finally develop.

**Calcium** is a major constituent of plant cell walls as calcium pectate which binds the cells together. It also influences the activity of **meristems**, especially in root tips. **Deficiency** causes weakened cell walls, resulting in inward curling, pale young leaves, and sometimes death of the growing point. Specific disorders produced include 'topple' in tulips, when the flower head cannot be supported by the top of the stem, 'blossom end rot' in tomato fruit, and 'bitter pit' in apple fruit.

**Sulphur** is a vital component of many enzymes, and is involved in the synthesis of **chlorophyll**. **Deficiency** produces chlorosis which, due to the relative immobility of sulphur, shows in younger leaves first.

**Iron** and **manganese** are involved in the synthesis of chlorophyll, although they do not form part of the molecule, and are components of some enzymes. **Deficiencies** of both minerals result in leaf chlorosis. The immobility of iron causes the younger leaves to show interveinal chlorosis first. In extreme cases, the growing area turns white.

**Boron** affects various processes, such as the translocation of sugars and the synthesis of gibberellic acid in some seeds (*see* dormancy). **Deficiency** causes a breakdown and disorganization of tissues, leading to early death of the growing point. Characteristic disorders include 'brown heart' of turnips, and 'hollow stem' in brassicas. The leaves may become misshapen, and stems may break. Flowering is often suppressed, while malformed fruit are produced, e.g. 'corky core' in apples, and 'cracked fruit' of peaches.

**Copper** is a component of a number of enzymes. **Deficiency** in many species results in dark green leaves, which become twisted and may prematurely wither.

**Zinc**, also involved in enzymes, produces characteristic deficiency symptoms associated with the poor development of leaves, e.g. 'little leaf' in citrus and peach, and 'rosette leaf' in apples.

## Mineral uptake

Minerals are absorbed in water by the root system, which obtains its supply from the growing medium (*see* Chapter 16). The movement of the elements, in the form of charged particles or **ions**, occurs in the direction of root cells containing a higher mineral concentration than the soil, i.e. against a **concentration gradient**. The passage in the water medium across the root cortex is by simple diffusion, but transport across the endodermis requires a supply of energy from the root cortex. The process is therefore related to temperature and oxygen supply (*see* respiration). The surface thickening which occurs in the ageing root does not significantly reduce the absorption ability of most minerals, e.g. potassium and phosphate, but calcium is found to be principally taken up by the young roots.

## PRACTICAL EXERCISES

1   Observe the effect of placing shoots of various plants in a solution containing 5 per cent **salt**.
2   Examine the cross-section of a carrot tap **root** as illustrated in Figure 4.1.
3   Observe the different rates of **water uptake** between a stem with leaves and a similar stem with leaves removed.
4   Examine leaves of various species, comparing their surface area and cuticle thickness.

## FURTHER READING

MAFF. *Diagnosis of Mineral Disorders in Plants*, Vol. 1 *Principles* (HMSO, 1984), Vol. 2 *Vegetables* (HMSO, 1984), Vol. 3 *Glasshouse Crops* (HMSO, 1987).

Richardson, M. *Translocation in Plants*, 2nd edition (Edward Arnold, 1971).

Scott Russell, R. *Plant Root Systems: Their Function and Interaction with the Soil* (McGraw-Hill, 1982).

Sutcliffe, J. *Plants and Water* (Edward Arnold, 1971).

Sutcliffe, J.F., and Baker, D.A. *Plant and Mineral Salts* (Edward Arnold, 1976).

# 5

# Plant Growth

*In any horticultural situation, a grower is concerned with controlling and even manipulating plant growth and development. He or she must provide for the plants the best conditions to produce the most efficient growth rate and the end product required. Therefore the processes which result in growth should be understood in order that the most economic growth results. Photosynthesis is probably the single most important process which needs to be provided for when growing crops. Respiration is the process by which the food matter produced by photosynthesis is converted into energy usable for growth of the plant.*

Growth is a difficult term to define because it really encompasses the totality of all the processes which take place during the life of an organism. However, it is useful to distinguish between the processes which result in an increase in size and weight, and those processes which cause the changes in the plant during its life cycle, which can usefully be called development, described in Chapter 6.

## PHOTOSYNTHESIS

**Photosynthesis is the process by which a green plant manufacturers food in the form of carbohydrates such as sugars and starch, using light as energy.** All living organisms require organic matter as food to build up their structure and to provide chemical energy to fuel their activities. Such complex organic compounds, based on carbon, must be produced from the simple raw materials, water and carbon dioxide. Many organisms are unable to manufacture their own food, and must therefore feed on already manufactured organic matter such as plants or animals. Since large animals predate on smaller animals, which themselves feed on plants, all organisms depend directly or indirectly on photosynthesis occurring in the plant as the basis of a **food web** or chain.

### Requirements for photosynthesis
### Carbon dioxide

In order that a plant may build up organic compounds, it must have a supply of carbon which is readily available. **Carbon dioxide** is present in the air in concentrations of 330 ppm (parts per million) or 0.03 per cent and can diffuse into the leaf through the stomata, as described in Chapter 4. Carbon dioxide gas moves ten thousand times faster than it would in solution through the roots. The amount of carbon dioxide in the air immediately surrounding the plant can fall when planting is very dense, or when plants have been photosynthesizing rapidly, especially in an unventilated greenhouse. This reduction will slow down the rate of photosynthesis, but a grower may supply additional carbon dioxide inside a greenhouse or polythene tunnel to **enrich** the atmosphere up to about three times the normal concentration, or an optimum of 1000 ppm (0.1 per cent) in lettuce. Such practices will produce a corresponding increase in growth, provided other factors are available to the plant. If any one is in short supply, then the process will be slowed down. **This principle, called the law of limiting factors, states that the factor in least supply will limit the rate of the process, and applies to other non-photosynthetic processes in the plant.**

It would be wasteful therefore to increase the carbon dioxide concentration artificially, e.g. by burning propane gas, or releasing pure carbon dioxide gas, if other factors were not proportionally increased.

### Light

Light is a factor required in order that photosynthesis can occur. In any series of chemical reactions where one substance combines with another to form a larger compound, energy is needed to fuel the reactions. Energy for photosynthesis is provided by light from the sun or from artificial lamps. As with carbon dioxide, the amount of light energy present is important in determining the rate of photosynthesis − simply, the more light or greater **illuminance** (intensity) absorbed by the plant, the more photosynthesis can take place. Light energy is measured in joules/square metre, but for practical purposes the light for plant growth is measured according to the light falling on a given area, that is lumens per square metre (lux). It is difficult to state the plant's precise requirements, as variation occurs with species, age, temperature, carbon dioxide levels, nutrient supply and health of the plant.

However, it is possible to suggest approximate limits within which photosynthesis will take place; a minimum intensity of about 500−1000 lux enables the plant's photosynthesis rate to keep

pace with **respiration**, and thus maintain itself. The maximum amount of light many plants can usefully absorb is approximately 30 000 lux, while good growth in many plants will occur at 10 000−15 000 lux. Plant species adapted to shade conditions, however, e.g. *Ficus benjamina*, require only 1000 lux. Other shade-tolerant plants include *Taxus spp.*, *Mahonia* and *Hedera*. In summer, light intensity can reach 50 000−90 000 lux and is therefore not limiting, but in winter months, between November and February, the low natural light intensity of about 3000−8000 lux is the limiting factor for plants actively growing in a heated greenhouse or polythene tunnel. Care must be taken to maintain clean glass or polythene, and to avoid condensation which restricts light transmission (Figure 5.1). Intensity can be increased by using artificial lighting, which can also extend the length of day, which is short during the winter, by **supplementary lighting**. This method is used for plants such as lettuce, bedding plants and brassica seedlings.

**Total replacement lighting**. Growing rooms which receive no natural sunlight at all use controlled temperatures, humidities, and carbon dioxide levels, as well as light. Young plants which can be grown in a relatively small area, and which are capable of responding well to good growing conditions in terms of growth rate, are often raised in a growing room.

**The type of lamp** chosen for increasing intensity, and therefore photosynthesis, is important. All such lamps must have a relatively high efficiency of conversion of electricity to light, and only the **gas discharge** lamps are able to do this. Light is produced when an electric arc is formed across the gas filament enclosed under pressure inside an inner tube. Light, like other forms of energy, e.g. heat, X-rays and radio waves, travels in the form of waves, and the distance between one wave peak and the next is termed the wavelength. Light wavelengths are measured in nanometres (nm); 1 nm = one thousandth of a micrometre. Visible light wavelengths vary from 800 nm (red light) to 350 nm (blue light), and a combination of different wavelengths (colours) appears as white light. Each

**Figure 5.1** *Condensation and dirt collected on the panes of a greenhouse. Both these factors can severely restrict the light entry into the greenhouse, especially on dull days.*

type of lamp produces a characteristic wavelength range and, just as different coloured substances absorb and reflect varying colours of light, so a plant absorbs and reflects specific wavelengths of light. Since the photosynthetic green pigment chlorophyll (*see* Chapter 6) absorbs mainly red and blue light and reflects more of the yellow and green part of the spectrum, it is important that the lamps used produce a balanced wavelength spectrum to include as high a proportion of those colours as possible, in order that the plant makes most efficient use of the light provided. The gas included in a lamp determines its light characteristics. The two most commonly used gases for horticultural lighting are mercury vapour,

producing a greeny-blue light with no red, and sodium, producing yellow light. This limited spectrum may be modified by the inclusion of fluorescent materials in the inner tube, which absorb long wavelengths emitted by the gas and re-emit the energy as a shorter wavelength. Thus, modified mercury lamps produce the desirable red light missing from the basic emission.

**Low pressure** mercury-filled tubes produce diffuse light and, when suitably grouped in banks, provide uniform light close to plants. These are especially useful in a growing room, provided that they produce a broad spectrum of light such as the full spectrum fluorescent tubes. Gas enclosed at **high pressure** in a second inner tube is a small, high intensity source of light. These small lamps do not greatly obstruct natural light entering a greenhouse and, while producing valuable uniform supplementary illumination at a distance, cause no leaf scorch. Probably the most useful lamp for supplementary lighting in a greenhouse is a high pressure sodium lamp which produces a high intensity of light, and is relatively efficient (27%).

Carbon dioxide **enrichment** should be matched to artificial lighting, in order to produce the greatest growth rate and most efficient use of both factors.

### Temperature

The complex chemical reactions which occur during the formation of carbohydrates from water and carbon dioxide require the presence of chemicals called **enzymes** to accelerate the rate of reactions. Without these enzymes, little chemical activity would occur. Enzyme activity increases with temperature from 0°C to 36°C, and ceases at 40°C. This pattern is very closely followed by the effect of temperature, both of the air and growing medium, on the rate of photosynthesis, depending on species, from 25°C to 36°C as optimum; but at very low light levels the increase in photosynthesis rate with increases in temperature is limited. This means that increased heat is wasted if the natural light is limiting. **Integrated environmental control** in a greenhouse requires continual monitoring, often by computer, of light, temperature and carbon dioxide so that heating and carbon dioxide requirements are correctly supplied.

The beneficial effects to plant growth of lower night temperatures compared with day are well known in many species, e.g. tomato. The explanation is inconclusive, but the accumulation of sugars during the night appears to be greater, suggesting a relationship between photosynthesis and respiration rates. Such responses are shown to be related to temperature regimes experienced in the origin areas of the species.

Adaptations to extremes in temperature can be found in a number of species; for example, resistance to high temperatures above 40°C in **thermophiles**; resistance to **chilling injury** is brought about by lowering the freezing point of cell constituents. Both depend on the stage of development of the plant, e.g. a seed is relatively resistant, but the hypocotyl of a young seedling is particularly vulnerable. Resistance to chilling injury is imparted by the cell membrane, which can also allow the accumulation of substances to prevent freezing of the cell contents. **Hardening-off** of plants by gradual exposure to cold temperatures can develop a change in the cell membrane, as in bedding plants and peas.

### Water

Water is required in the photosynthesis reaction, but this represents only a very small proportion of the total water taken up by the plant (*see* transpiration). Water supply through the xylem is essential to maintain leaf turgidity and retain fully open stomata for carbon dioxide movement into the leaf. Due to shortage of carbon dioxide, a 10 per cent loss of water may lead to a 50 per cent decrease in photosynthesis, while a visibly wilting plant will not be photosynthesizing at all.

### Minerals

Minerals are required by the leaf to produce the **chlorophyll** pigment which absorbs most of the light energy for photosynthesis. Production of chlorophyll must be continuous, since it loses its efficiency quickly. A plant deficient in iron, or magnesium especially, turns yellow (**chlorotic**) and loses much of its photosynthetic ability. **Variegation** similarly results in a slower growth rate.

### The chloroplast

This is the subcellular unit for photosynthesis

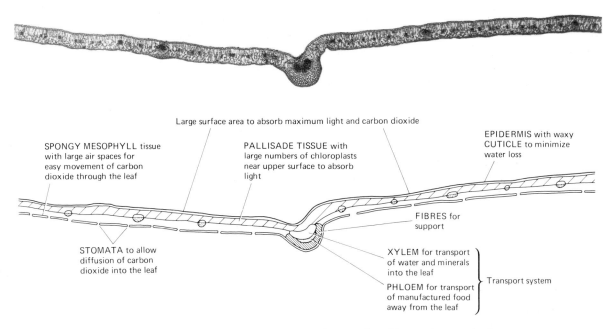

**Figure 5.2** *Cross-section of privet* (Ligustrum) *leaf showing features for efficient photosynthesis.*

which draws together the raw materials described. The absorption of light by chlorophyll occurs at one site and the energy is transferred to a second site where it is used to build up carbohydrates, usually in the form of insoluble starch.

### The leaf

The leaf is the main organ for photosynthesis in the plant, and its cells are organized in a way that provides maximum efficiency. Figure 5.2 shows the structure of the leaf and its relevance to the process of photosynthesis. A newly expanded leaf is most efficient in the absorption of light, and this ability reduces with age.

**Pollution.** Gases in the air, which are usually products of industrial processes or burning fuels, can cause damage to plants, often resulting in scorching symptoms of the leaves. Fluoride can accumulate in composts and be present in tap water, so causing marginal and tip scorch in leaves

of susceptible species such as *Dracaena* and *Gladiolus*.

Sulphur dioxide and carbon dioxide may be produced by faulty heat exchangers in glasshouse burners, especially using paraffin. Scorch damage over the whole leaf is preceded by a reddish discoloration.

### RESPIRATION

The product of photosynthesis in most plants is starch (some plants produce sugars only), which is stored temporarily in the chloroplast or moved in the phloem to be more permanently stored in the seed, the stem cortex or root, where specialized storage organs such as **rhizomes** and **tubers** may occur. The movement or **translocation** of materials around the plant in the phloem and xylem is a complex operation, and does not have a full scientific explanation. The phloem is principally responsible for the transport of the products of photosynthesis as soluble sugars which move under pressure to areas of need, such as roots, flowers or storage organs. Energy is needed to maintain the movement, but whether to 'fuel' the activity or maintain the phloem structure is not understood. The flow can be interrupted by the presence of disease organisms such as club root.

**Respiration is the process by which sugars and related substances are broken down to yield energy, the end-products being carbon dioxide and water.** In order that growth can occur, the food must be broken down in a controlled manner to release energy for the production of useful structural substances such as cellulose, the main constituent of plant cell walls, and proteins for **enzymes**. This energy is used also to fuel cell division and the many chemical reactions which occur in the cell. The energy requirement within the plant varies, and reproductive organs can respire at twice the rate of the leaves. **In order that the breakdown is complete, oxygen is required in the process of aerobic respiration. In the absence of oxygen, inefficient anaerobic respiration takes place and incomplete breakdown of the carbohydrates produces alcohol as a waste product, with energy still trapped in the molecule.** If a plant or plant organ such as a root is supplied with low oxygen concentrations in a waterlogged or compacted soil, the consequent alcohol production may prove toxic enough to cause root death. Overwatering, especially of pot plants, leads to this damage and encourages damping-off fungi.

The process of photosynthesis and respiration constitute an **energy cycle** where light energy is trapped, stored, and finally converted to chemical energy for growth. A plant which is adequately supplied with heat for photosynthesis will subsequently respire efficiently and no special provision need be made in the growing of the crop.

### Storage of plant material

The actively growing plant is supplied with the necessary factors for photosynthesis and respir-

ation to take place. Roots, leaves or flower stems removed from the plant for sale or planting will cease to photosynthesize, though respiration continues. Carbohydrates and other storage products, such as proteins and fats, continue to be broken down to release energy, but the plant reserves are depleted and dry weight reduced. A reduction in the respiration rate should therefore be considered for stored plant material, whether the period of storage is a few days, e.g. tomatoes and cut flowers, or several months, e.g. apples. Attention to the following factors may achieve this aim.

**Temperature**. The enzymes involved in respiration become progressively less active with a reduction in temperatures from 36°C (optimum) to 0°C.

Therefore, a cold store employing temperatures between 0°C and 10°C is commonly used for the storage of materials such as cut flowers, e.g. roses; fruit, e.g. apples; vegetables, e.g. onions; and cuttings, e.g. chrysanthemums, which root more readily later. Long term storage of seeds in **gene banks** (*see* Chapter 7) uses liquid nitrogen at −20°C.

**Oxygen and carbon dioxide**. Respiration requires oxygen in sufficient concentration; if oxygen concentration is reduced, the rate of respiration will decrease. Conversely, carbon dioxide is a product of the process and, as with many processes, a build-up of a product will cause the rate of the process to decrease. A controlled environment store for long term storage, e.g. of top fruit, is maintained at 0°C−5°C according to cultivar, and is fed with inert nitrogen gas to exclude oxygen. Carbon dioxide is increased by up to 10% for some apple cultivars.

**Water loss** may quickly desiccate and kill stored material, such as cuttings. Seeds also must not be allowed to lose so much water that they become non-viable, but too humid an environment may encourage premature germination with equal loss of viability.

## PRACTICAL EXERCISES

1  In a commercial greenhouse, examine the systems for **controlling** temperatures and carbon dioxide.
2  Examine types of **lamps** used for supplementary lighting in a greenhouse, and total replacement lighting in a growing room.
3  Observe relative growth rates for plants subjected to high and low light **intensities**.
4  Observe germination rates of seed **stored** under various conditions: for example, wet, cold, dry, hot.

## FURTHER READING

Bickford, E.D., and Dunn, S. *Lighting for Plant Growth* (Kent State University Press, 1973).

Electricity Council. Grow Electric Handbook No. 2 *Lighting in Greenhouses*, Part 1 (1974).

Bleasdale, J.K.A. *Plant Physiology in Relation to Horticulture* (Macmillan, 1983).

MAFF. *Carbon Dioxide Enrichment for Lettuce*, HPG51 (HMSO, 1978).

Richardson, M. *Translocation in Plants*, 2nd edition (Edward Arnold, 1975).

Sutcliffe, J. *Plants and Temperature* (Edward Arnold, 1977).

# 6

# Plant Development

*The life cycle of the flowering plant contains a number of identifiable stages, each with a distinct significance to horticulture. The seed is the means by which a new generation, often of variable plants, begins; the sensitive seedling, vulnerable to diseases, pest attack and physiological disorders, is highly responsive to good growing conditions; the vegetative stage may be manipulated to the required size and shape or used for propagation; the flowering stage is often the desired objective, while the formation of fruit may be an important horticultural aim, whether in a succulent form or as the precursor of seeds.*

The characteristics of the plant are under genetical control (*see* Chapter 7). **However, the growth and development of the plant, from one stage of the life cycle to the next, takes place in response to a number of stimuli, many of which are changes in factors in the plant's environment.** The plant's most reliable indicator of season, especially in temperate regions, is the **daylength**, which shortens and lengthens with the time of year. Annual temperature fluctuations are a valuable, if less precise, indicator of seasonal occurrence. The plant responds to these environmental factors by breaking seed or bud dormancy in spring, by flowering at appropriate seasons, and by dropping (**abscission**) of leaves in autumn.

In order that the genetical potential can be stimulated by factors of the environment, the plant must possess a system which, having perceived the stimulus, is able to activate the response. Such control is achieved by chemicals produced by the plant in quantities creating a specific balance depending on the stage of development and the response required. These chemicals are grouped into the **auxins**, involved in stimulating cell division and enlargement, the **gibberellins**, which particularly control stem extension growth and dormancy, and the **cytokinins**, which are important in stimulating cell division.

Other substances, for example, **abscisic acid** and ethylene, do not belong to any particular chemical group but, by responses such as accelerating growth, breaking dormancy, and inducing flowering, are able to contribute to the precision characteristic of the **hormonal system**.

The horticulturist is able to both modify the plant environment and also utilize chemicals similar in action to those produced by the plant, called **plant growth regulators**, in order to manipulate the growth and development of the plant. The relevance of these applications will be discussed at each stage of development.

## THE SEED

The seed, resulting from sexual reproduction,

creates a new generation of plants which bear characteristics of both parents. The plant must survive often through conditions which would be damaging to a growing vegetative organism, and the seed is a means of protecting against extreme conditions of temperature and moisture, as the form for **overwintering**. The seed structure may be specialized for wind dispersal, e.g. members of the Compositae family, including groundsel, dandelion and thistle, have parachutes, as does clematis (Ranunculaceae). Many woody species such as lime (*Tilia*), ash (*Fraxinus*), and sycamore (*Acer*) produce winged fruit. Other seedpods are explosive, e.g. balsam and hairy bittercress. Other organisms such as birds and mammals distribute hooked fruits such as goosegrass and burdock, and succulent (e.g. tomato, blackberry, elderberry) or protein filled fruit, for example, dock.

### Seed structure (*see* Figure 6.1)

A seed, in order to survive, must contain a small immature plant (**embryo**) protected by a seed coat or **testa**, which is formed from the outer layers of the ovule after **fertilization**. A weakness in the testa, the **micropyle**, marks the point of entry of the pollen tube prior to fertilization, and the

**hilum** the point of attachment to the fruit. The embryo consists of a **radicle**, which will develop into the primary root of the seedling, and a **plumule**, which develops into the shoot system, the two being joined by a region called the **hypocotyl**. A single seed leaf (**cotyledon**) will be found in monocotyledons, while two are present as part of the embryo of dicotyledons. The cotyledons may occupy a large part of the seed, e.g. in beans, to act as the food store for the embryo. In some species, e.g. grasses and castor oil, the food of the seed (**endosperm**) is derived from the fusion of extra nuclei at the time of fertilization. Plant food is usually stored as the carbohydrate, starch, formed from sugars as the seed matures, for example in peas and beans. Other seeds, such as sunflowers, contain high proportions of fats and oils, and proteins are often present in varying proportions.

### Seed dormancy

As soon as the embryo begins to grow out of the seed, i.e. **germinates**, the plant is vulnerable to damage from cold or drought. Therefore, the seed must have a mechanism to prevent germination when poor growing conditions prevail. **Dormancy is a period during which very little activity occurs in the seed, other than a very slow rate of respiration.**

A **thick testa** prevents water and oxygen, essential in germination, from entering the seed. Gradual breakdown of the testa, occurring through bacterial action or freezing and thawing, eventually permits germination after the unsuitable conditions. The passing of fruit through the digestive system of an animal such as a bird may promote germination, for example in tomato, cotoneaster, holly. Many species, e.g. fat hen, produce seed with variable dormancy periods, to spread germination time over a number of growing seasons. Spring soil cultivations can break the seed coat and induce germination of weed seeds. This structural dormancy, in horticultural crops, may present germination problems in plants such

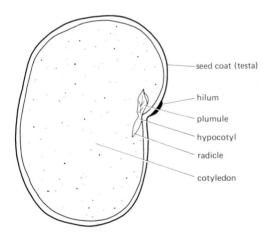

seed coat (testa)

hilum

plumule

hypocotyl

radicle

cotyledon

**Figure 6.1** *Long section of a broad bean seed showing the structure of seed coat and embryo essential for germination.*

as rose rootstock species and *Acacia*. Physical methods, using sandpaper or chemical treatment with sulphuric acid (collectively known as **scarification**), can break down the seed coat and therefore the dormancy mechanism.

**Chemical inhibitors** may occur in the seed to prevent the germination process. **Abscisic acid** at high concentrations helps maintain dormancy while, as dormancy breaks, progressively lower levels occur, with a simultaneous increase in concentrations of growth promotors such as gibberellic acid and cytokinins. Inhibitory chemicals located just below the testa may be washed out by soaking in water. Cold temperatures cause similar breaks of dormancy, **stratification**, in other species, the exact temperature requirement varying with the period of exposure and the plant species. Many alpine plants require a 4°C stratification temperature, while other species, e.g. *Ailanthus*, *Thuja*, ash, and many other trees and shrubs, require both moisture and the chilling treatment. The chemical balance inside the seed may be changed in favour of germination by treatment with chemicals such as gibberellic acid and potassium nitrate.

An **undeveloped embryo** in a seed is incapable of germinating until time has elapsed after the seed is removed from the parent plant, i.e. the **after-ripening** period has occurred, as in the tomato, and many tropical species such as palms.

### Seed germination

**Seed germination is the emergence of the radicle through the testa, usually at the micropyle.** There are a number of essential germination requirements in order that successful seedling emergence occurs.

A **viable seed** has the potential for germination given the required external conditions. Its viability, therefore, indicates the activity of the seed's internal organs, i.e. whether the seed is 'alive' or not. Most seeds remain viable until the next growing season, a period of about eight months, but many can remain dormant for a number of years until conditions are favourable for germination. In general, viability of a batch of seed diminishes with time, its maximum viability period depending largely on the species. For example, celery seed quickly loses viability after the first season, but wheat has been reported to have germinated after scores of years. The germination potential of any seed batch will depend on the storage conditions of the seed, which should be cool and dry, slowing down respiration, and maintaining the internal status of the seed. These conditions are achieved in commercial seed stores by means of sensitive control equipment. Packaging of seed for sale takes account of these requirements and often includes a waterproof lining of the packet, which maintains a constant water content in the seeds.

**Water supply** to the seed is the first environmental requirement for germination. The water content of the seed may fall to 10 per cent during storage, but must be restored to about 70 per cent to enable full chemical activity. Water initially is absorbed into the structure of the testa in a way similar to a sponge taking up water into its air space, i.e. by **imbibition**. This softens the testa and moistens the cell walls of the seed in order that the next stages can proceed. The cells of the seed take up water by **osmosis**, and provide a suitable medium for the activity of enzymes in the process of respiration. A continuous water supply is now required if germination is to proceed at a consistent rate, but the growing medium, whether it is outdoor soil or compost in a seed tray, must not be waterlogged, because **oxygen** essential for aerobic respiration would be withheld from the growing embryo. In the absence of oxygen, **anaerobic respiration** occurs and eventually causes death of the germinating seed, or suspended germination, i.e. **induced dormancy**.

**Temperature** is a very important germination requirement, and is usually specific to a given species or even cultivar. It acts by fundamentally influencing the activity of the enzymes involved in the biochemical processes of respiration which occur between 0°C and 40°C. However, species adapted to specialized environments respond to a narrow range of germination temperatures. For example, cucumbers require a minimum temperature of 15°C, and tomatoes 10°C. On the other hand, lettuce germination may be inhibited by temperatures higher than 30°C, and in some cultivars, at 25°C, a period of induced dormancy occurs. Some species, such as mustard, will germinate in temperatures just above freezing and up to 40°C, provided they are not allowed to dry out.

**Light** is a factor which may influence germination in some species, but most species are indifferent. Seed of *Rhododendron*, *Veronica* and *Phlox* is inhibited in its germination by exposure to light, while that of celery, lettuce, most grasses, conifers and many herbaceous flowering plants is slowed down when light is excluded. This should be taken into account when the covering material for a seed bed is considered (*see* tilth).

The colour (wavelength) of light involved may be critical in the particular response created. Far red light (720 nm), occurring between red light and infra-red light and invisible to the human eye, is found to inhibit germination in some seeds, e.g. birch, while red light (660 nm) promotes it. A canopy of tall deciduous plants filters out red light for photosynthesis. Seed of species growing under this canopy receive mainly far red light, and are prevented from germinating. When the leaves fall in autumn, these seeds will germinate both in response to the now available red light and to the low winter temperatures.

### The Seeds Acts

In the UK the Seeds Acts control the quality of seed to be used by growers. A seed producer must satisfy the minimum requirements for species of vegetables and forest tree seed by subjecting a seed batch to a government testing procedure. A sample of the seed is subjected to standardized ideal germination conditions, to find the proportion which is viable (**germination percentage**). The germination and emergence under less ideal field conditions (**field emergence**) where tilth and disease factors are variable, may be much lower than germination percentage. The sample is also tested for **quality** which provides information, available to the purchaser of the seed, covering trueness to type — that is, whether the characteristics of the plants are consistent with those of the named cultivar; the percentage of non-seed material, such as dust; the percentage of weed seeds, particularly those of a poisonous nature (*see* **Weeds Act**, Chapter 12). The precise regulations for sampling and testing, and requirements for specific species, have changed slightly since the 1920 Act, the 1964 Act (which also included the details of plant varieties), and entry of Britain into the European Community. Some control under EC regulations is made of the provenance of forestry seed, as the geographical location of its source is important in relation to a number of factors, including response to drought, cold, dormancy, habit, and pest and disease susceptibility.

### THE SEEDLING

Within the seed is a food store which provides the means to produce energy for germination. Once the food store has been exhausted, the seedling must rapidly become independent in its food supply and begin to photosynthesize. **It must therefore respond to stimuli in its environment to establish the direction of growth. Such a response is termed a tropism**, and is very important in the early survival of the seedling (Figure 6.2).

**Geotropism** is a directional growth response to gravity. The emergence of the radicle from the testa is followed by growth of the root system, which must quickly take up water and minerals in order that the shoot system may develop. A seed germinating near the surface of a growing medium must not put out roots which grow on to the surface and dry out, but the roots must grow downwards to tap water supplies. Conversely, the plumule must grow away from the pull of gravity in order that the leaves develop in the light.

**Etiolation** is the type of growth which the shoot

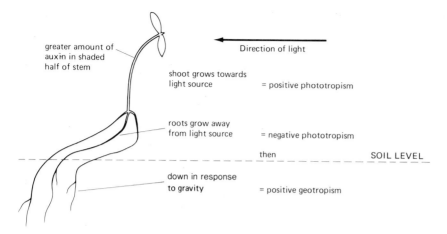

greater amount of auxin in shaded half of stem

Direction of light

shoot grows towards light source

= positive phototropism

roots grow away from light source

= negative phototropism

then                              SOIL LEVEL

down in response to gravity

= positive geotropism

**Figure 6.2** *Geotropism and phototropism shown as mechanisms assisting the survival of a seedling.*

produces as it moves through the soil in response to gravity. The developing shoot is delicate and vulnerable to physical damage, and therefore often the growing tip is protected by being bent into a plumular hook. The stem grows quickly, is supported by the structure of the soil and therefore is very thin and spindly, stimulated by friction in the soil which causes release of ethylene. The leaves are undeveloped, as they do not begin to function until they move into the light. Mature plants which are grown in dark conditions also appear etiolated.

### Seedling development

The emergence of the plumule above the growing medium is usually the first occasion that the seedling is subjected to light. This stimulus inhibits the extension growth of the stem so that it becomes thicker and stronger, but the seedling is still very susceptible to attack from pests and damping-off diseases. The leaves unfold and become green in response to light, which enables the seedling to photosynthesize and so support itself. The first leaves to develop, the cotyledons, derive from the seed and may emerge from the testa while still in the soil, as in peach and broad bean (**hypogeal** germination), or be carried with the testa into the

air, where the cotyledons then expand (**epigeal** germination), e.g. in tomatoes and cherry.

**Phototropism** occurs in order that the shoot grows towards a light source which provides the energy for photosynthesis. A bend takes place in the stem just below the tip as cells in the stem away from the light grow larger than those near to the light source. A greater concentration of auxin in the shaded part of the stem causes the extended growth (Figure 6.2). Roots display a negative phototropic response, growing away from light when exposed at the surface of the growing medium, for example, on a steep bank. The growth away from light may supersede the root's geotropic response, and will cause the roots to grow back into the growing medium.

**Hydrotropism** is the growing of roots towards a source of water. The explanation of this tropism has not been found, but it can be shown to occur.

The **cotyledons** which emerge from the testa contribute to the growth of the seedling in photosynthesis, but the **true leaves** of the plant, which often have a different appearance to the cotyledon, very quickly unfold.

### THE VEGETATIVE PLANT

The role of the vegetative stage in the life cycle of

the plant is to grow rapidly and establish the individual in competition with others. It must therefore photosynthesize effectively and be capable of responding to good growing conditions. Growing rooms with near-ideal conditions of light, temperature, and carbon dioxide utilize this capacity which will reduce with the ageing of the plant (*see* Chapter 5).

## Juvenility

The early growth stage of the plant, the **juvenile** growth, is characterized by certain physical appearances and activities which are different to those found in the later stages or in **adult** growth. Often leaf shapes vary, as shown in Figure 6.3: the juvenile ivy leaf is three-lobed, while the adult leaf is oval. Other examples are common in conifer species, where the complete appearance of the plant is altered by the change in leaf form, for example, *Chamaecyparis fletcheri* and many

*Juniperus* species such as *J. chinensis*. In the genera *Chamaecyparis* and *Thuja*, the juvenile condition can be fixed by repeated vegetative propagation producing a plant called **retinospores**, which are used as decorative features.

**Leaf retention** is also a characteristic of juvenility (*see* Figure 6.4). It can be significant in species such as beech, where the phenomenon is exaggerated, as an additional protection in **wind breaks** where the trees can be pruned back to the vegetative growth. The habit of the plant is also different, as are its activities. The juvenile stem of ivy is horizontally growing and vegetative in nature, while the adult growth is vertical and bears flowers. Therefore, the juvenile stage is a period after germination which is capable of rapid vegetative growth and is unlikely to flower. Many species which require an environmental change to stimulate flower initiation, such as the Brassicas which require a cold period, will not respond to the stimulus until the juvenile period is over − about eleven weeks in Brussels sprouts.

**Figure 6.3** *Adult growth (left) of ivy showing entire leaves and flowers; juvenile growth (right) showing typical ivy-shaped leaf and adventitious roots.*

**Figure 6.4** *Leaf retention in the lower juvenile branches of Beech* (Fagus).

The adult stage essential for sexual reproduction is less useful for vegetative propagation than the responsive juvenile growth, a condition probably due to the hormonal balance in the tissues. Figure 6.3 shows the spontaneous production of adventitious roots on the ivy stem. Adult growth should be removed from stock plants to leave the more successful juvenile growth for cutting.

**Vegetative propagation**

Although the life cycle of most plants leads to sexual reproduction, **all plants have the potential to reproduce asexually or by vegetative propagation, when pieces of the parent plant are removed and develop into a wholly independent plant**. All living cells contain a nucleus with a complete set of genetical information (*see* genetic code, Chapter 7), with the potential to become any specialized cell type. Only part of the total information is brought into operation at any one time and position in the plant. If parts of the plant are removed, then cells lose their orientation in the whole plant and are able to produce organs in positions not found in the usual organization. These are described as **adventitious** and can, for example, be roots on a stem cutting, buds on a piece of root, or roots and buds on a piece of leaf used for vegetative propagation.

Many plant species use the ability for vegetative propagation in their normal pattern of development, in order to increase the number of individuals of the species in the population. The production of these vegetative **propagules**, as with the production of seed, is often the means by which the plant survives adverse conditions (*see* overwintering), acting as a food store which will provide for the renewed growth when it begins. The stored energy in the swollen tap roots of dock and dandelion enable these plants to compete more effectively with seedlings of other weed and crop species, which would also apply to roots of carrots and beetroot. Stems are telescoped in the form of a **corm** in freesia and cyclamen, or swollen into a **tuber** in potato or cyclamen, or a **rhizome** in iris and couch grass. Leaves expanded with food may form a large bud or **offset** found in lilies, or a **bulb** as in onions, daffodils and tulips (Figure 6.5). Other natural means of propagation include lateral stems, which grow horizontally on the soil surface to produce nodal adventitious roots and subsequently plantlets, i.e. **runners** or **stolons** of strawberries and yarrow. The adventitious nature of stems is exploited when they are deliberately bent to touch the ground, or enclosed in compost, in the method known as **layering**, used in carnations, some apple rootstocks, many deciduous shrubs, and pot plants such as *Ficus* and *Dieffenbachia*.

The roots of species, especially in the Rosaceae family, are able to produce adventitious buds which grow into aerial stems or **suckers**, e.g. pears, raspberries, and many nursery stock species. By all these methods of runners, layering and suckers, the propagule will subsequently become detached from the parent plant by the disintegration of the connecting stem or root.

All these natural methods are used in horti-culture to produce numbers of plants from a single parent plant. This group of plants, or **clone**, strictly speaking, is an extension of the parent plant, and therefore all have the same genetical characteristics. The horticulturist is able to reproduce a cultivar precisely by this means, whereas seed production is likely to result in **variation** of characteristics. However, changes can occur (*see* mutations), and differing clonal characteristics within the same cultivar can be distinguished in some conifer species, e.g. the leaf colour and plant habit of × *Cupressocyparis leylandii*.

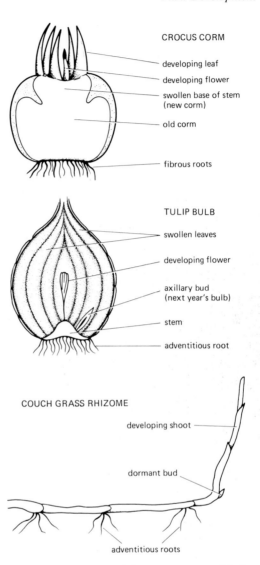

CROCUS CORM
- developing leaf
- developing flower
- swollen base of stem (new corm)
- old corm
- fibrous roots

TULIP BULB
- swollen leaves
- developing flower
- axillary bud (next year's bulb)
- stem
- adventitious root

COUCH GRASS RHIZOME
- developing shoot
- dormant bud
- adventitious roots

**Figure 6.5** *Structure of the organs responsible for overwintering and vegetative propagation.*

Seeds produced without fertilization by **apomixis** found in Rosaceae and Graminae present a special case of natural clonal propagation.

The artificial methods of vegetative propagation utilize all organs of the plant.

**Stem cuttings** may be of several types of stem material. Hardwood cuttings are pieces of dormant woody stem containing a number of buds which grow out into shoots when dormancy is

**Figure 6.6** *Tissue cultures of* Sorghum bicolour *showing: A – early growth of callus tissue, and B – later plantlet production.*

broken in spring. The base of the cutting is cut cleanly to expose the cambium tissue from which the adventitious roots will grow (e.g. in rose rootstocks, *Forsythia*, and many deciduous ornamental shrubs). In *Hydrangea* and currant the root initials are pre-formed. Semi-ripe cuttings have a woody **heel**, which produces the roots, and a herbaceous shoot system which will develop further after rooting (e.g. in holly and many conifers). Stems without a woody nature are used for the propagation of *Fuchsias*, *Pelargoniums*, *Chrysanthemums*, and many others. The area of

leaf on these cuttings should be kept to a minimum to reduce water loss. This risk can further be reduced by **misting**, that is spraying the plants with fine droplets of water to increase humidity and reduce temperature, with a consequent slowing of transpiration rate. Automatic misting employs a switch attached to a sensitive device used for assessing the evaporation rate from the leaves. The cool conditions favouring the survival of the aerial parts of the cutting do not encourage the division of cells in the cambium area of the root initials. Therefore, the temperature in the rooting medium may be increased with electric cables producing **bottom heat**. These special conditions for the encouragement of the success of cuttings are provided in propagation benches in a greenhouse.

**Leaf cuttings** are also susceptible to wilting before the essential roots have been formed, and will benefit from mist, provided the wet conditions do not encourage rotting of the plant material. Leaves of species such as *Begonia*, *Streptocarpus* and *Sansevieria* are divided into pieces from which small plantlets are initiated, while the leaves plus petioles are used for *Saintpaulia* propagation. Nursery stock species, e.g. *Camelia* and Rhododendron, require a complete leaf and associated axillary bud in a leaf-bud cutting.

**Tissue culture is a method used for vegetative propagation, employing a small piece of plant tissue, the explant, grown in a sterile artificial medium supplying all mineral and organic nutrients, and enclosed in a vessel which is subjected to precisely controlled environmental conditions** (Figures 6.6 and 6.7). This method has advantages over conventional propagation techniques, since a valuable parent stock plant may be used to produce large numbers of propagules if small pieces of tissue are used, and the time of bulking up stock is considerably reduced. Some species which traditionally propagate only by seed, e.g. orchids and asparagus, can now be grown by tissue culture. One of the problems of vegetative propagation is that diseases and pests are passed on to the propagules. Stock plants, encouraged by high temperatures to grow quickly, produce

**Figure 6.7** *Plantlets growing in tubs under tissue culture conditions in a commercial laboratory.*

low levels of disease (particularly virus) in their growing tips. The **meristem-tip** can be dissected out of the stem and grown in a tissue culture medium, to produce stock free from disease (e.g. chrysanthemum stunt viroid, *see* Chapter 11). This method of propagation is now used for species including *Begonia*, *Alstroemeria*, *Figus*, *Malus*, *Pelargonium*, Boston fern (*Nephrolepsis exaltata*), roses and many others.

In all the methods described, **cell division** (*see* mitosis), must be stimulated in order to produce the new tissues and organs. This **initiation** is triggered by the balance of hormones produced by the cells. Auxins are found to stimulate the initiation of adventitious roots of cuttings, and the bases of cuttings may be dipped in powder or liquid formulations of auxin-like chemicals such as naphthalene acetic acid. The number of roots is increased and production time reduced. The precise concentration of chemical in the cells is critical in producing the desired growth response. A large amount of hormone can bring about an inhibition of growth rather than promotion. Different organs respond to different concentration ranges; for example, the amount of auxin needed to increase stem growth would inhibit the production of roots. The same principle applies to another group of chemicals important in cell division, the cytokinins, which can be applied to leaf cuttings to increase the incidence of plantlet formation. Both chemicals must be included in a tissue culture medium, at concentrations appropriate to the species and the type of growth required. The weaning of plantlets from their protected environment in tissue culture conditions requires care and usually conditions of high relative humidity, shade and warmth.

### Apical dominance

After the germination of the seed, the plumule establishes a direction of growth due partly to the geotropic and phototropic forces acting on it. Often the terminal bud of the main stem sustains the major growth pattern, while the axillary buds are inhibited in growth to a degree which depends on the species. In tomatoes and chrysanthemums, the lateral shoots will grow out, but are inhibited by a high concentration of auxin which accumulates in these buds. The source of the chemical is the terminal bud which sustains a promotive concentration. Removal of the main shoot, or **stopping**, takes away the supply to the axillary buds which are then able to grow out to a degree, dependent on the exact concentration retained.

The practice of **disbudding**, as in chrysanthemums and carnations, takes out the axillary buds to allow the terminal bud to develop into a bigger bloom because of greater food availability.

Parts of plants can be removed or **pruned** to reduce the competition within the plant for the available resources. So, a reduction in the number of flower buds of, for example, chrysanthemum will cause the remaining buds to develop into larger flowers; a reduction in fruiting buds of apple trees will produce bigger apples, and the reduction in branches of soft fruit and ornamental shrubs will allow the plants to grow stronger when planted densely. Pruning will also obviously affect the shape of the plant, as meristems previously inhibited by apical dominance will begin to develop. The success of such pruning depends very much on the skill of the operator, as a good knowledge of the species habit is required.

### Growth retardation

Stem extension growth is controlled by auxins produced by the plant, and also by gibberellins which can dramatically increase stem length, especially when externally applied. Growth retardation may be desirable, especially in the production of compact pot plants from species which would normally have long stems, for example, chrysanthemums, tulips, *Azaleas*. Therefore, artificial chemicals such as daminozide or phosfon (Figure 6.8), which inhibit the action of the growth promoting hormones, can retard the development of the main stem, and also stimulate the growth of side shoots to produce a more bushy, compact plant. Flower production may be inhibited, but this can be countered by the application of flower stimulating chemicals.

## THE FLOWERING PLANT

The progression from a vegetative to a flowering plant involves profound physical and chemical changes. The stem apex displays a more complex appearance under the microscope as flower initiation occurs, and is followed, usually irreversibly, by the development of a flower. The stimulus for this change may simply be genetically derived, but often an environmental stimulus is required which links flowering to an appropriate season.

### Photoperiodism

**Photoperiodism is a daylength stimulated response involved in dormancy in buds and seeds, leaf abscission, and the initiation of flowers.** This last process is probably the most important for manipulation by the horticulturist. Repeated daily exposure to light for a given period of time (**critical period**), the length of which depends on the species, will bring about flower initiation, provided the light periods are separated by an appropriate period of darkness. Some species are adapted to flower after exposure to daylengths shorter than a critical period. These are **short day plants**, and adaptation will ensure that they produce flowers at the most favourable time for successful seed production and survival. *Chrysanthemum morifolium*, *Poinsettia* and *Kalanchoe* species must have short days followed by relatively long nights in order that flower buds are produced. **Long day plants** which require a daylength longer than a critical period include radish, *Nigella* and carnation. Although this latter species will flower under any photoperiodic conditions, flowers are produced more readily when treated with long days. Many species previously considered to be non-responsive or **day neutral**, e.g. *Begonia elatior* and tomato, may respond in some degree to a lighting period, especially in combination with temperature. Plants such as greenhouse chrysanthemums can be artificially induced to flower even though the natural daylength conditions are not suitable. The photoperiod can be extended by artificial lighting, or shortened by blacking out the daylight. The relatively low light intensity of approximately 100–150 lux, needed to stimulate

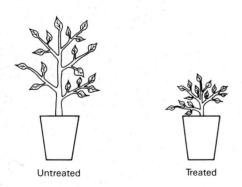

Untreated          Treated

**Figure 6.8** *Chemical growth retardant is incorporated into compost used for pot plants such as chrysanthemums.*

a response, is adequately provided by incandescent tungsten filament lamps placed above the crop, while blackout is provided using black polythene (Figure 6.9). In a short day plant such as chrysanthemum, the night period must be relatively long in order to bring about a flowering response. If a period of artificial lighting divides the night into two periods, i.e. a **night-break**, then the flowering response does not occur. This practice is used to maintain vegetative growth in order to produce a full length stem at a time when a natural short day would cause the plants to flower prematurely. Light is absorbed in the photoperiodic response by a blue pigment called **phytochrome** which is in insufficient quantities to

contribute colour to the plant. It absorbs principally red light from the spectrum, which represents a considerable proportion of the emission of tungsten filament bulbs. The stimulus of the phytochrome by red light occurs in a few seconds, but can be reversed by far red light which is just outside the visible light spectrum. This exposure to far red light will prevent the response expected, i.e. flower initiation. The same reversal is caused by the dark period, and the number of dark or light periods therefore contributes to the response. Since the reversal by a dark period is much slower than by pure far red light, and the plant is insensitive to dark periods shorter than fifteen minutes, a repeated lighting procedure of fifteen minutes

**Figure 6.9** *Chrysanthemum blooms with lighting and blackout for daylength control.*

light, followed by fifteen minutes dark (i.e. **cyclic lighting**) represents continuous light to the plant.

### Flower initiation

Flower initiation can be stimulated largely by photoperiodic or temperature changes, or a complex interaction between temperature and daylength. Cold temperatures experienced during the winter bring about flower initiation (i.e. **vernalization**) in many biennial species such as *Brassica*, lettuce, red beet, *Lunaria* and onion. The period for the response depends on the exact temperature, as with bud-break and seed dormancy (*see* stratification). The optimum temperature for many of these responses is about 4°C. Hormones are involved in causing the flower apex to be produced. The balance of auxins, gibberellins and cytokinins is important, but some species respond to artificial treatment of one type of chemical; for example, the daylength requirement for chrysanthemum plants can be partly replaced by gibberellic acid sprays.

### Flower structure (*see* Figure 6.10)

The flower is initially protected inside a flower bud by the calyx or **sepals**, which are often green and can therefore photosynthesize. The development of the flower parts requires a large energy expenditure by the plant, and therefore vegetative activities decrease. The corolla or **petals** may be small and insignificant in wind-pollinated flowers, e.g. grasses, or large and colourful in insect-pollinated species. The colours and size of petals can be improved in cultivated plants by breeding, and may also involve the multiplication of the petals or **petalody**, when fewer male organs are produced. **The flowers of many species have both male and female organs (hermaphrodite), but some have separate male and female flowers (monoecious), e.g. *Cucurbita*, walnut, birch (*Betula*) whereas others produce male and female flowers on different plants (dioecious), e.g. holly, willows,**

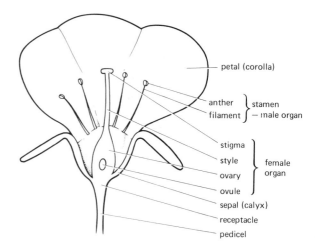

**Figure 6.10** *Cross-section of a flower showing the structures involved in the function of sexual reproduction.*

*Skimmia japonica* **and** *Ginkgo biloba*. Each male **stamen** bears an anther which produces and discharges the **pollen grains**. Identification of present and past genera is possible by microscopic examination of pollen. The female organ is positioned in the centre of the flower and consists of an ovary containing one or more **ovules** or egg cells. The style leads from the ovary to a stigma at its top where pollen is captured. The flower parts are positioned on the **receptacle** which is the tip of the flower stalk or **pedicel**. Associated with the flower head or **inflorescence** are leaf-like structures called **bracts**, which can sometimes assume the function of insect attraction, e.g. in *Poinsettia*.

### Flower development

The flower opens to expose the organs for sexual reproduction. The life of the flower is limited to the time needed for pollination and fertilization, but it is often commercially desirable to extend the life of a cut flower or flowering pot plant. In cut flowers, water uptake must be maintained and dissolved nutrients for opening the flower bud are termed an **opening solution**. A sterilant, e.g. silver nitrate, in the water can reduce the risk of blockage of xylem by bacterial or fungal growth.

**Ethylene** has a considerable effect on flower development, and can bring about premature death (**senescence**) of the flower after the flower begins to open. Cut flowers should therefore never be stored near to fruit, e.g. apples, which produce ethylene. Some chemicals, such as sodium thiosulphate, reduce the production of ethylene in carnations and therefore extend their life. Vase life can be extended by the addition of sterilants and sugar to the water.

### Pollination and fertilization

The function of the flower is to bring about sexual reproduction, which is the production of offspring following the fusion of male and female nuclei. The male and female nuclei are contained within the pollen grain and ovule respectively, and **pollination is the transference of the pollen from the anther to the stigma**. The source of pollen may be the same flower in **self-pollination** or a different flower in **cross-pollination**, although for practical plant breeding purposes cross-pollination should include different plants. Cross-pollination ensures that variation is introduced into new generations of offspring. Natural agents of cross-pollination are mainly wind and insects.

**Wind-pollinated** flowers are small, green, lacking nectar and unscented, but produce large amounts of pollen, for example willow (*Salix*). **Insect pollinated** flowers produce odours and have brightly-coloured petals to attract insects and produce nectar to feed them. The insects then collect pollen on their bodies and carry it to other flowers. **Bees** are important pollinating insects in orchards, where cross-pollination is essential, as in some triploid cultivars such as Bramley's seedling. Plant breeders may use house flies, enclosed in a glasshouse, to carry out pollination, or mechanical transference of pollen may be achieved by hand. Some floral mechanisms, for example snapdragon, clover, physically prevent non-pollinating species by allowing only heavy bees to enter the flower. Others trap pollinators for a period of time to give the best chance of successful fertiliz-

ation, e.g. *Arum* lily. Certain *Primula spp.* have stigma and stamens of differing lengths to ensure cross-pollination.

The pollen grain on arrival at the stigma produces a pollen tube, after absorption of sugar and moisture from the stigma's surface. The male nuclei are carried in the pollen tube, which grows down inside the style and into the ovary wall. After entering the ovule, **male and female nuclei fuse together in fertilization, resulting in the zygote which will divide and differentiate to become the embryo of the seed**. The seed endosperm may be produced by more nuclear fusion, and the testa of the seed forms from the outer layers of the ovule. The ovary develops into a protective fruit.

### THE FRUITING PLANT

The development of the fruit involves either the expansion of the ovary into a juicy **succulent** structure, or the tissues becoming hard and **dry**, both providing protection and/or dispersal of the seeds. The succulent fruits are often eaten by animals which helps seed dispersal, and may also bring about chemical changes to break dormancy mechanisms. The dry fruits may rot away gradually to release the seeds (**indehiscent**), or may burst open to release the seed (**dehiscent**). Such mechanisms are found in fruits of the Compositae family with tiny parachutes for wind dispersal (e.g. groundsel). In Leguminosae the pods or **legumes** split open, and in some Crucifers and poppies the seeds are released quite violently. This can account for the widespread distribution of weed species such as hairy bittercress.

### Fruit set

Fruit set is stimulated by the process of pollination in most species. The hormones, in particular gibberellins, carried in the pollen, trigger the production of auxin in the ovary which causes the cells to develop. In species such as cucumber, the naturally high content of auxin enables fruit production without prior fertilization, i.e. **parthenocarpy**, a useful phenomenon when the object of the crop is the production of seedless fruit. Such activity can be simulated in other species, especially when poor conditions of light and temperature have caused poor fruit set in species such as tomato and peppers. Here, the flowers are sprayed with an auxin-like chemical; however, the quality of fruit is inferior. Pears can be sprayed with a solution of gibberellic acid to replace the need for pollination. Fruit ripening occurs as a result of hormonal changes and involves in tomatoes a change in the sugar content, i.e. at **climacteric**. After this point, fruit will continue to ripen and also respire after removal from the plant. Ethylene is released by ripening fruit, which contributes to deterioration in store. Early ripening can be brought about by a spray of a chemical, e.g. ethephon, which stimulates the release of ethylene by the plant, e.g. in the tomato.

## THE AGEING PLANT

At the end of an annual plant's life, or the growing season of perennial plants, a number of changes take place. The changes in colour associated with autumn are due to pigments which develops in the leaves and stems.

**Pigments** are substances which are capable of absorbing light; they also reflect certain wavelengths of light which determine the colour of the pigment. Chlorophyll, which reflects mainly green light, is produced in considerable amounts, and therefore the plant, especially the leaves, appears predominantly green. Other pigments are present; for example, the carotenoids (yellow) and xanthophylls (red), but usually the quantities are so small as to be masked by the chlorophyll. In some species, for example copper beech (*Fagus sylvatica*), other pigments predominate, masking chlorophyll. These pigments also occur in many species of deciduous plants at the end of the growing season, when chlorophyll synthesis ceases prior to the abscission of the leaves. Many colours

are displayed in the leaves at this time in such species as *Acer platanoides*, turning gold and red, *Prunus cerasifera* 'Pissardii' with light purple leaves, European larch with yellow leaves, Virginia creeper (*Parthenocissus* and *Vitus spp.*) with red leaves, beech with brown leaves, *Cotoneaster* and *Pyracantha* with coloured berries, and *Cornus* species, which have coloured stems. These are used in **autumn colour** displays at a time when few flowering plants are seen out of doors.

In deciduous woody species, the leaves drop in the process of **abscission**, which may be triggered by shortening of the daylength. In order to reduce risk of water loss from the remaining leaf scar, a corky layer is formed before the leaf falls. Auxin production in the leaf is reduced, and this stimulates the formation of the abscission layer, and abscisic acid is involved in the process. Premature leaf fall in nursery stock plants can be brought about by auxin spray. Ethylene inhibits the action of auxin, and can therefore also cause premature leaf fall, for example, in *Hydrangea* prior to cold treatment for flower initiation.

## PRACTICAL EXERCISES

1   Collect a number of seeds of different species, e.g. broad bean, and sunflower; cut them open to examine the **structure** illustrated in Figure 6.1.
2   Examine the effects of varying temperatures on the **germination** of cucumber seed.
3   Observe the **etiolated** growth of plants subjected to low light intensities; and **phototropism** in plants where light falls from one direction, e.g. tomato seedlings, or carnations growing in a greenhouse in winter.
4   Search for plants with **juvenile** and **adult** leaf form.
5   Cut across a tulip, daffodil or hyacinth bulb, a freesia or crocus corm, and examine the internal **structure** as illustrated in Figure 6.5.
6   In a commercial greenhouse producing all-year-round chrysanthemums, investigate the

relative lighting and blackout periods for different seasons.

7   Collect flowers from a suitable plant, e.g. rose or *Hypericum*; remove the parts and examine the **structure**, as illustrated in Figure 6.10.

8   In autumn search for species displaying various aspects of **autumn colour**.

## FURTHER READING

Bleasdale, J.K.A. *Plant Physiology in Relation to Horticulture* (Macmillan, 1983).

Electricity Council. Grow Electric Handbook No. 2 *Lighting in Greenhouses*, Part 2, 1974.

Hart, J.W. *Plant Tropisms and Other Growth Movements* (Unwin Hyman, 1990).

Leopold, A.C. *Plant Growth and Development*, 2nd edition (McGraw-Hill, 1977).

MAFF. *Quality in Seeds*, Advisory Leaflet 247.

MAFF. *Guide to the Seed Regulations* (HMSO, 1979).

Wareing, P.F., and Phillips, I.D.J. *The Control of Growth and Differentiation in Plants* (Pergamon, 1981).

Wilkins, M.B. *Physiology of Plant Growth and Development* (McGraw-Hill, 1969).

# 7

# Genetics and Plant Breeding

*Ever since man selected seed for his next crop, he has influenced the genetical make-up and therefore the potential of succeeding crops. A basic understanding of plant breeding principles enables the grower to understand the potential and limits of plant cultivars and therefore make more realistic requests to the plant breeder. Plant breeding now supplies a wide range of plant types to meet growers' specific needs. The plant breeder's skill relies on his knowledge of genetics to manipulate inheritable characters, and his ability to recognize and select the desirable characters when they occur.*

**Plant breeding uses the basic principle that fundamental characteristics of a species are passed on from one generation to the next (heredity).** However, sexual reproduction may produce different characteristics in the offspring (**variation**). A plant breeder relies on the principles of heredity to retain desirable characteristics in a breeding programme, and new characteristics are introduced in several ways to produce new cultivars.

In order to understand the principles of plant breeding, the genetical make-up of the plant must first be studied.

## THE CELL

Every living plant cell contains a **nucleus** which controls every activity occurring in the cell. Within the nucleus is the chemical deoxyribose nucleic acid (DNA), a very large molecule made up of thousands of atoms, the arrangement of which can vary. This specific arrangement determines the genetic information provided by the DNA (**genetic code**), and a change in the code dictates new characteristics of the plant. The number of these long chain-like molecules of DNA (**chromosomes**) varies with the plant species. The cells of the tomato (*Lycopersicum esculentum*), contain twenty-four chromosomes, the cells of *Pinus* and *Abies* species twenty-four, onions sixteen (human beings have forty-six). Each chromosome contains a succession of units, called **genes**, each of which usually codes for a single characteristic. Scientists have been able to relate the many gene locations to plant characteristics. Microscopic observation of cells during cell division reveals two similar sets of chromosomes, e.g. in tomatoes, twelve different pairs. This condition is termed **diploid**. A gene for a particular characteristic, such as flower colour, has a precise location on one chromosome, and on the same location of the paired chromosome. For each characteristic, therefore, there are at least two genes, one on each chromosome in the pair, which may provide genetical information for that characteristic.

### Cell division

When a plant grows, the cell number increases in the growing points of the stems and roots, the division of one cell producing two new ones. **Genetic information in the nucleus must be reproduced exactly in the new cells to maintain the plant's characteristics. This is achieved by the process of mitosis, where each chromosome in the parent cell produces a duplicate of itself, sufficient material for the two new daughter cells.** A delicate, spindle-shaped structure ensures the separation of chromosomes, one complete set into each of the new cells. A dividing cell wall forms across the old cell to complete the division.

Sexual reproduction involves the fusion of genetical material contained in the sex cells from both parents (*see* fertilization). Half the chromosomes in the cells of an offspring are therefore inherited from the male parent, and half from the female. To ensure that the offspring chromosome number equals that of the parents, contents of the nuclei of male and female sex cells (*see* pollen and ovule) must be halved. **The cell division process (meiosis), occurring in the anthers and ovaries, resulting in the nuclei of the sex cells ensures the separation of each chromosome from its partner so that each sex cell contains only one complete set of chromosomes. This condition is termed haploid.**

## INHERITANCE OF CHARACTERISTICS

Genetic information is passed from parent to offspring when material from male and female parent comes together by fusion of the sex cells. Genes from each parent can, in combination, produce a mixture of the parents' characteristics in the offspring; for example, a gene for red flowers inherited from the male parent, combined with a gene for white flowers from the female

parent, could produce pink-flowered offspring. If one of the genes, however, completely dominated the other, for example if the red gene inherited from the male parent was **dominant** over the white female gene, all offspring would produce red flowers. The **recessive** white gene will still be present as part of the genetic make-up of the

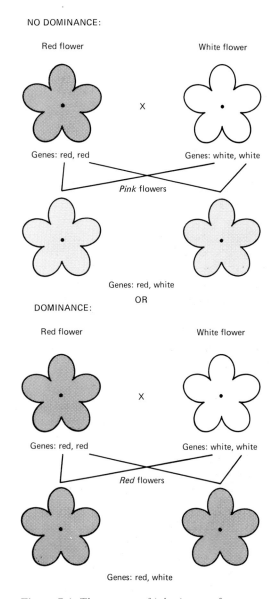

**Figure 7.1** *The pattern of inheritance of genes.*

offspring cells and can be passed on to the next generation. If it then combines at fertilization with another white gene, the offspring will be white-flowered (Figure 7.1).

The example given considers the inheritance of one pair of genes by a single offspring. Many sex cells are produced from one flower in the form of pollen grains and ovules, which fuse to form many seeds of the next generation. The plant breeder must know the genetic composition of the whole population of offspring. Consider now the same example of flower colour. If a red-flowered plant containing two genes for red (described as **pure**), is crossed with a pure white flowered plant, the red flowered plant supplies pollen as the male parent, and the white flowered plant produces seed as the female parent. As both parents are pure, the male parent can produce only one type of sex cell, containing 'red' genes, and the female parent only ovules with genes for 'white' flowers. Since all pollen grains will carry 'red' genes and all ovules 'white', then in the absence of dominance, the only possible combination for the first generation or $F_1$ is pink offspring, each containing a gene for red and a gene for white (i.e. **impure**). Figure 7.2 illustrates this inheritance by using letters to describe genes, **R** to represent a red gene, and r to represent a white gene.

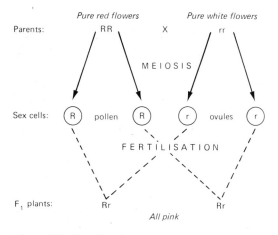

**Figure 7.2** *Simple inheritance.*

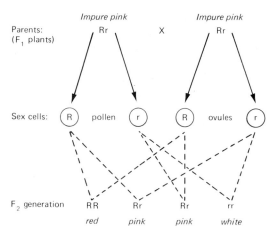

**Figure 7.3** *Simple inheritance.*

If the plants from the $F_1$ generation were used as parents and crossed (or perhaps self-pollinated), then the results in the second or $F_2$ generation would be as shown in Figure 7.3: 25 per cent of the population, therefore, would have red flowers, 50 per cent pink flowers, and 25 per cent white flowers.

The plant breeder, analysing the ratios of each colour, would therefore be able to calculate which colour genes were present in the parents, and whether the 'colour' was pure.

## $F_1$ HYBRIDS

**The importance of $F_1$ hybrids to the grower is that, given a uniform environment, plants of the same cultivar which are $F_1$ hybrids will produce a uniform crop because they are all genetically identical** (Figure 7.2). Crops grown from $F_1$ hybrid seed, such as cabbage, Brussels sprouts and carrots, can be harvested at one time, and have similar characteristics of yield; flower crops will have uniformity of colour and flower size. Plants derived from crossing parent types with different characteristics display **hybrid vigour**, and generally respond well to good growing conditions. The desirable characteristics of the two parents, such as disease resistance, good plant habit, high yield, and good fruit or flower quality, may be incorporated along with established characteristics of successful commercial cultivars by means of the $F_1$ hybrid breeding programme.

$F_1$ hybrid seed production first requires suitable parent stock, which must be pure for all characteristics. In this way, genetically identical offspring are produced, as described in Figure 7.2. The production of pure parent plants involves repeated self-pollination (**selfing**) and selection, over eight to twelve generations, resulting in suitable **inbred parent lines**. During this and other self-pollination programmes vigour is lost but, of course, is restored by hybridization.

The parent lines must now be cross-pollinated to produce the $F_1$ hybrid seed. It is essential to avoid self-pollination at this stage, therefore one of the lines is designated the male parent to supply pollen. The anthers in the flowers of the other line, the female parent, are removed, or treated to prevent the production of viable pollen. The growing area must be isolated to exclude foreign pollen, and seed is collected only from the female parent. This seed is expensive compared with other commercial seed, due to the complex programme requiring intensive labour.

Seed collected from the commercial $F_1$ hybrid crop represents the $F_2$, and will produce plants with very diverse characteristics (Figure 7.3). It is clear, therefore, that seed from an $F_1$ should not be saved for sowing, if uniform characteristics need to be retained. Some $F_2$ seed is produced in flowering plants such as geraniums and fuchsias, where variable colour and habit are required.

## OTHER BREEDING PROGRAMMES

In addition to $F_1$ hybrid breeding, where specific improvements are achieved, plant breeders may wish to bring about more general improvements to existing cultivars, or introduce characteristics such as disease resistance. Programmes are required for crops which self-pollinate (**inbreeders**), or those which cross-pollinate (**outbreeders**), and three of these strategies are described.

*Pedigree breeding* is the method for plant and animal breeding most widely used by both amateurs and professionals. Two plants with different desirable characteristics are crossed to produce an $F_1$ population. These are selfed and the offspring with useful characteristics are selected for further selfing to produce a line of plants. After repeated selfing and selection, the characteristics of the new lines are compared with existing cultivars and assessed for improvements. Further field trials will determine a new type's suitability for submission and potential registration as a new cultivar.

If a plant breeder wishes to produce a strain adapted to particular conditions, such as hardiness, exposure of plants of a selected cultivar to the desired conditions will eliminate unsuitable plants, thus allowing the hardy plants to set seed. Repetition of this process gradually adapts the whole population, while other characteristics, such as earliness, may be selected by harvesting seed early.

**Disease resistance** (*see* Chapter 12), a genetic characteristic, enables the plant to combat fungal attack. The disease organism may itself develop a corresponding genetical capacity to overcome the plant's resistance by mutation. The introduction of disease resistance into existing cultivars requires a **backcross breeding** programme, involving a cultivar lacking resistance, and often a wild species with resistance. A lettuce cultivar, for example, may lack resistance to downy mildew (*Bremia lactucae*), or a tomato cultivar lack tomato mosaic virus resistance. This commercial cultivar is crossed with a resistant wild plant to produce an $F_1$, then an $F_2$. From this $F_2$, plants having both the characteristics of the commercial cultivar and also disease resistance are selected. The process continues with backcrossing of these selected plants with the original commercial parent to produce an $F_1$, from which an $F_2$ is produced. More commercial characteristics may be incorporated by further backcrossing and selection over a number of generations, until all the characteristics of the commercial cultivar are restored, but with the additional disease resistance.

## POLYPLOIDS

Polyploids are plants with cells containing more than the diploid number of chromosomes; for example, a triploid has three times the haploid number, a tetraploid four times, and the polyploid series continues in many species up to octaploid (eight times haploid). An increase in size of cells, with a resultant increase in roots, fruit and flower size of many species of chrysanthemums, fuchsias, strawberries, turnips and grasses, is the result of polyploidy, although a maximum chromosome number sustained by any one species can be identified. Polyploidy occurs when duplication of chromosomes (*see* mitosis), fails to result in mitotic cell division. The multiplication of a polyploid cell within a meristem may form a complete polyploid shoot which, after flowering, may produce polyploid seed. The crossing of a tetraploid and a diploid gives rise to a triploid. Polyploid plants are often infertile, especially **triploid**, with an odd number of chromosomes which are unable to pair up during meiosis. A number of apple cultivars, such as Bramley's Seedling and Crispin, are triploid and require pollinator cultivars for the supply of viable pollen. Although polyploidy can occur spontaneously, and has led to many variant types in wild plant populations, it can be induced by the use of a mitosis inhibitor, colchicine.

## MUTATIONS

Changes in the content or arrangement of the chromosomes cause changes in the characteristics of the individual. Very drastic alterations result in malformed and useless plants, while slight rearrangements may provide a horticulturally desirable change in flower colour or plant habit, as shown in chrysanthemums, dahlias or *Streptocarpus*. Mutation breeding has produced these variations using irradiation treatments with X-rays, gamma rays, or mutagenic chemicals. Spontaneous mutations regularly occur in cells but, as in polyploids, the mutation only becomes significant in the plant when the mutated cell is part of a meristem.

A shoot with a different colour flower or leaf may arise in a group of plants, and is termed a **sport**. When a plant consists of two or more genetically distinct tissues, it is called a **chimaera**, often resulting in variegation of the leaves, e.g. *Acer* and *Pelargonium*. These conditions are maintained only by vegetative propagation, which encourages cell division in both tissue types. All mutations are inherited by a succeeding generation in the ways previously described, which enables new characteristics and potential new cultivars to be produced in just one generation.

The **Plant Varieties and Seeds Act, 1964** protects the rights of producers of new cultivars. The registration of a new cultivar is acceptable only when its characteristics are shown to be significantly different from any existing type. Successful registration enables the plant breeder to control the licence for the cultivar's propagation, whether by seed or vegetative methods. Separate schemes operate for the individual genera of horticultural and agricultural crops, but all breeding activities may benefit from the 1964 Act.

### Gene banks

As new, specialized cultivars are produced and grown, using highly controlled cultivation methods, old cultivars and wild sources of variation – which could be useful in future breeding programmes – are lost. Therefore, a gene bank provides a means of storing the seed of many cultivars and species at very low temperatures, while some plant material is maintained in tissue culture conditions. One such bank is situated at the National Vegetable Research Station, Wellesbourne, Warwickshire, England.

## PRACTICAL EXERCISES

1  If an impure, red-flowered plant is crossed with a white-flowered plant, and red is domi-

nant to white, what would be the flower colours of the resulting $F_1$ plants?

2 Look through seed catalogues for examples of $F_1$ hybrids, $F_2$ populations, triploid and tetraploids.

3 Investigate the origins of cultivated plants of interest, and search for reference to the breeding principles described in this chapter.

## FURTHER READING

Have, D.J. van der. *Plant Breeding Perspectives* (Centre for Agricultural Publishing and Documentation, Wageningen, 1979).

North, C. *Plant Breeding and Genetics in Horticulture* (Macmillan, 1979).

Simmonds, N.W. *Principles of Crop Improvement* (Longman, 1979).

# 8

# Weeds

> *Most plants, whether wild or cultivated, grow in competition with other organisms such as pests, diseases, and other plants. Any competitive plant unwanted by man is termed a weed. This chapter indicates the growing situations where weeds become problems, the means of weed identification, the specific biological features of weeds that make them important, and the relevant control measures used against them. Detailed principles of control are described in Chapter 12.*
>
> *A **weed** is a plant of any kind which is growing in an undesirable place: groundsel smothering lettuce, moss covering a lawn, last year's potatoes emerging in a plot of cabbage, rose suckers spoiling a herbaceous border.*

## IDENTIFICATION

As with any problem in horticulture, recognition and identification are essential before any control measures can be attempted. The weed seedling causes little damage to a crop and is relatively easy to control. Identification of this stage is therefore important and, with a little practice, the grower may learn to recognize the important weeds using such features as shape, colour, and hairiness of the cotyledons and first true leaves (*see* Figure 8.1).

Early recognition of a weed species may be crucial for control. The ivy-leaved speedwell is susceptible to the foliage-applied herbicides only at the cotyledon to three-leaf stages, and attempted control against the more mature weed will give poor results.

Within any crop or bedding display, a range of different weed species will be observed. Changes in the weed flora may occur because of environmental factors such as reduced pH; because of new crops which may encourage different weeds to develop; or because repeated use of one herbicide selectively encourages certain weeds, e.g. groundsel in lettuce crops or annual meadow grass in turf. Horticulturists must watch carefully for these changes so that their chemical control may be adjusted.

The mature weeds may be identified using an illustrated 'flora' which shows details of leaf and flower characters.

**Damage** caused by weeds may be categorized into seven main areas.

1  **Competition** between the weed and the planting for water, nutrients and light may prove favourable to the weed if it is able to establish itself quickly. The plants are therefore deprived of their major requirement and poor growth results. The extent of this competition is largely unpredictable, varying with climatic factors such as temperature and rainfall, soil

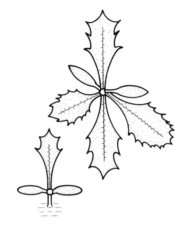

**Chickweed** (x 1.5)
Bright green. Cotyledons have a light coloured tip and a prominent mid-vein. True leaves have long hairs on their petioles

**Groundsel** (x 1.5)
Cotyledons are narrow and purple underneath. True leaves have step-like teeth

**Large Field Speedwell** (x 1.5)
Cotyledons like the 'spade' on playing cards. True leaves hairy, notched and opposite

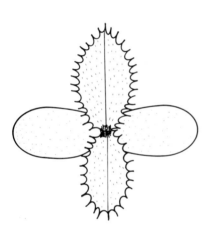

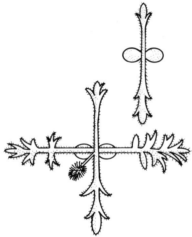

**Creeping Thistle** (x 1.5)
Cotyledons large and fleshy. True leaves have prickly margins

**Yarrow** (x 1.5)
Small broad cotyledons. True leaves hairy and with pointed lateral lobes

**Broad-leaved Dock** (x 1.5)
Cotyledons narrow. First leaves often crimson, rounded with small lobes at the bottom

**Figure 8.1** *Seedlings of common weeds. Notice the difference between cotyledons and true leaves. (Reproduced by permission of Blackwell Scientific Publications.)*

factors such as soil type, and cultural factors such as cultivation method, plant spacing and quality of weed control in previous seasons. Large numbers of weed seed may be introduced into a plot in poor quality composts or farmyard manure. The uncontrolled proliferation of weeds will inevitably produce serious plant losses.

2 **Seed quality** is lowered by weed seed, e.g. fat hen contamination in batches of seed, e.g. carrots.

3 **Machinery**, e.g. mowing machines and harvesting equipment may be fouled by weeds such as knotgrass, which have stringy stems.

4 **Poisonous plants**. Ragwort, sorrel and buttercups are eaten by farm animals when more

desirable food is scarce. Poisonous fruits, e.g. black nightshade, may contaminate mechanically harvested crops such as peas for freezing.

5  **Pests and diseases** are commonly harboured on weeds. Chickweed supports whitefly, red spider mite and cucumber mosaic virus in greenhouses. Sowthistles are commonly attacked by chrysanthemum leaf miner. Groundsel is everywhere infected by a rust which attacks cinerarias. Charlock may support high levels of club root, a serious disease of brassica crops. Fat hen and docks allow early infestations of black bean aphid to build up. Speedwells may be infested with stem and bulb nematodes.

6  **Drainage** (*see* Chapter 14) depends on a free flow of water along ditches, Dense growth of grasses, e.g. couch, and water-loving weeds, e.g. chickweed, may seriously reduce this flow and increase waterlogging of horticultural land.

7  **Tidiness** is essential in a well-maintained garden. The amenity horticulturist may consider that any plant spoiling the appearance of plants in pots, borders, paths or lawns should be removed, even though the garden plants themselves are not affected.

### Weed biology

The range of weed species includes algae, mosses, liverworts, ferns (bracken) and flowering plants. These species display one or more special features of their life cycle which enable them to compete as **successful** weeds against the crop and cause problems for the horticulturist.

Many weeds, e.g. groundsel, produce seeds which germinate more quickly than crop seeds and thus emerge from the soil to crowd out the developing plants. Seeds of species such as charlock, annual meadow grass and groundsel germinate throughout the year, while others such as orache appear in spring, and cleavers in autumn. Carefully timed cultivations or herbicidal control is effective against this growth stage.

The soil conditions may favour certain weeds. Sheep's sorrel prefers acid conditions, while mosses are found in badly drained soils. Knapweed competes well in dry soils, while common sorrel survives well on phosphate-deficient land. Yorkshire fog grass invades poorly fertilized turf, while nettle and chickweed favour highly fertile soils.

The growth habit of the weed may influence its success. Chickweed and slender speedwell produce prostrate stems bearing numerous leaves which prevent light reaching emergent plant seedlings, while groundsel and fat hen have an upright habit which competes less for light in the early period. Weeds such as bindweed, cleavers and nightshades are able to grow alongside woody plants or soft fruit and border shrubs, making control difficult.

Different weeds propagate in different ways. Chickweed completing several life cycles per year (**ephemeral**), and black nightshade completing one (**annual**), both produce seed in order to continue the species. Annual seed production may be high in certain species. A scentless mayweed plant may produce 300 000 seeds, fat hen 70 000, and groundsel 1000. A **dormancy** period is seen in many weed species, groundsel being an exception. Seed germination commonly continues over a period of four or five years after seed dispersal, presenting the grower with a continual problem.

Some of the most difficult perennial weeds to control rely on **vegetative propagation** for their long-term survival in soils. Bracken and couch grass survive and spread by means of underground stems (rhizomes). Field bindweed and creeping thistle, on the other hand, produce creeping roots which in the latter species may grow six metres in one year. Docks, dandelions, plantains develop swollen tap roots, while horsetails survive the winter by means of underground tubers. The large quantities of food stored in vegetative organs enable these species to emerge quickly from the soil in spring, often from considerable depths if the soil has been ploughed in. The fragmentation of underground rhizomes and creeping roots by cultivation machinery enables these species to increase in disturbed soils. Weeds with swollen roots provide the greatest problems to the horticulturist in long-term crops such as soft fruit and turf because **foliage-acting** and residual herbicides may have little effect.

Fragmentation of above-ground parts may be important. A lawnmower used on turf containing the slender speedwell weed cuts and spreads the delicate stems which, under damp conditions, establish on other parts of the lawn.

**Control** measures are regularly necessary in most crop and amenity situations. Greenhouse production suffers much less from weed problems because composts and border soils are regularly sterilized.

Below are described important annual and perennial weeds. Specific descriptions of identification, damage, biology and control measures are given for each weed species.

Detailed discussion of weed control measures (**legislative**, **cultural**, and **chemical**) is presented in Chapter 12.

## ANNUAL WEEDS

Three species, chickweed, groundsel and speedwell, are described below to demonstrate some features of their biology that make them successful weeds.

### Chickweed (*Stellaria media*)

This species is found in many horticultural situations as a weed of flower beds, vegetables, soft fruit, and greenhouse plantings. It has a wide distribution throughout Britain, grows on land up to altitudes of 700 m, and is most often seen on rich, heavy soils.

The seedling cotyledons are pointed, with a light coloured tip, while its true leaves have hairy petioles (Figure 8.1). The adult plant has a characteristic lush appearance (Figure 8.2) and grows in a prostrate manner over the surface of the soil; in some cases it covers an area of $0.1 \, m^2$, its leafy

**Figure 8.2** *Chickweed: note the opposite leaves and small white flowers.*

stems crowding out young plants as it increases in size. Small white, five-petalled flowers are produced throughout the year, the flowering response being indifferent to day-length. The flowers are self-fertile. An average of 2500 disc-like seeds (1 mm in diameter) may result from the oblong fruit capsules produced by one plant. Since the first seed may be dispersed within six weeks of the plant germinating, and the plant continues to produce seed for several months, it can be seen just how prolific the species is. The seeds are normally released as the fruit capsule opens during dry weather; they survive digestion by animals and birds and may thus be dispersed over large distances. Irrigation water may carry them into channels and ditches. The large numbers of seed (up to fourteen million per hectare) are most commonly found in the top 7 cm of the soil where, under conditions of light, fluctuating temperatures and nitrate ions, they may overcome the **dormancy** mechanism and germinate to form the seedling. Many seeds, however, survive up to the second, third, and occasionally fourth years. Figure 8.3 shows that germination can occur at any time of the year, with April and September as peak periods.

Chickweed is an **alternate host** for many aphid transmitted viruses (e.g. cucumber mosaic), and the stem and bulb nematode.

Control of this weed is best achieved by a combination of methods. Partial sterilization of soil in greenhouses is effective, while hoeing in the spring and autumn periods prevents the developing seedling from flowering. The weed may be killed by **pre-emergent contact** sprays, e.g. paraquat applied before a crop emerges, by **soil applied** root acting herbicides, e.g. propachlor, or by **foliage-acting** herbicides applied in specific crop situations, e.g. linuron in potatoes. The weed in greenhouses may, with care, be controlled by

carefully directed sprays of contact herbicides, e.g. pentanochlor.

### Groundsel (*Senecio vulgaris*)

This is a very common and important weed, found in many countries, particularly on heavy soil. It grows on both poor and rich soils up to almost 600 m in altitude.

The seedling cotyledons are narrow, purple underneath, and the true leaves have step-like teeth (Figure 8.1). The adult plant has an upright habit, and produces as many as twenty-six yellow, small-petalled flower heads; flowering occurs in *all* seasons of the year. About forty-five column-shaped seeds, 2 mm in length, and densely packed in the fruit head, bear a mass of fine hairs which, when released in dry weather, carry the individual seeds along on air currents for many metres, while in wet weather the seeds become **sticky** and may be carried on the feet of animals, including humans. The seeds survive digestion by birds, and thus can be transported in this way. As can be seen in Figure 8.3, the seeds may germinate at any time of the year, with early May and September as peak periods. Since there may be more than three generations of groundsel per year (the autumn generation surviving the winter), and each generation may give rise to a thousand seeds, it is clear why groundsel is one of the most successful colonizers of cultivated ground. Its role as a symptomless carrier of the wilt fungus *Verticillium* increases its importance in certain crops, e.g. hops.

A combination of control methods may be necessary for successful control. Hoeing or alternative cultivation, particularly in spring and autumn, prevents developing seedlings from flowering, but uprooted flowering groundsel plants do produce viable seed. **Contact** herbicides, e.g. paraquat, may be sprayed to control the weed on paths or in fallow soil. **Soil-acting** chemicals, e.g. propachlor on brassicas, kill off the germinating seedling. An established groundsel population, especially in a crop such as lettuce, a

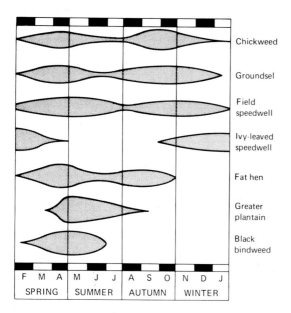

**Figure 8.3** *Annual and perennial weeds: periods of seed germination. Note that chickweed, groundsel and field speedwell seeds germinate throughout the year. Many other species are more limited. (Reproduced by permission of Blackwell Scientific Publications.)*

fellow member of the Compositae family, requires careful choice of herbicide to avoid damage to the crop.

**Speedwells** (*Veronica persica* and *V. filiformis*)

The first species, the large field speedwell, is an important weed in vegetable production, while the second species, the slender or round-leaved speedwell, is no longer considered a pleasant rock garden plant but a serious turf problem.

The seedling cotyledons are spade-shaped, while the true leaves are opposite, notched and hairy (Figure 8.1) in both species. The adult plants have erect, hairy stems, and rather similar broad, toothed leaves.

*V. persica* produces up to 300 bright blue flowers, 1 cm wide, per plant. The flowers are self-fertile and occur throughout the year, but mainly between February and November. The adult plant produces an average of 2000 light brown boat-shaped seeds (2 mm across), which fall to the ground and may be dispersed by ants; the rather large seeds travel as contaminants of crop seeds. The seeds of this species germinate below soil level all year round, but most commonly in March to May (Figure 8.3), the winter period being necessary to break **dormancy**. Seeds may remain viable for more than two years.

*V. filiformis* produces **self-sterile** purplish-blue flowers between March and May, and spreads by means of prostrate stems which root at their nodes to invade fine and coarse turf, especially in damp areas. Segments of this weed cut by lawnmowers easily root and further disperse the species. Seeds are not important in its spread.

Control of *V. persica* is best achieved by a combination of methods. Hoeing or alternative cultivation, particularly in spring, prevents developing seedlings from growing to mature plants and producing their many seeds. **Contact** herbicides, e.g. paraquat, may be sprayed to control the weed on paths or in fallow soils. **Soil-acting** chemicals, e.g. chlorpropham on onions, kill off the germinating weed seedling, while **contact**

chemicals, e.g. clopyralid, may be sprayed on to a young onion crop to control emerging seedlings.

The slender speedwell (*V. filiformis*) represents a different problem for control. While preventative control on turf seedbed with a **contact** chemical, e.g. paraquat, allows the turf to establish undisturbed, use of a more **selective** contact chemical, e.g. ioxynil, is necessary to prevent chemical damage to established grass. Regular close mowing and spiking of turf removes the high humidity necessary for this weed's establishment and development.

While there are at least fifty successful annual weed species in horticulture, this book can cover only a few examples which illustrate the main points of life cycles and control.

## PERENNIAL WEEDS

Four species, creeping thistle, couch, yarrow, and broad-leaved dock are described below to demonstrate the major features of their biology (particularly perennating organs) which make them successful weeds.

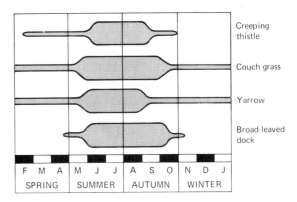

**Figure 8.4** *Perennial weeds: periods of flowering. Most flowers and seeds are produced between June and October. Annual weeds commonly flower throughout the year. However the slender speedwell flowers only between March and May.*

**Creeping thistle** (*Cirsium arvense*)

This species is a common weed in grassland and perennial crops, e.g. apples, where it forms dense clumps of foliage.

The seedling cotyledons are broad and smooth, the true leaves spiky (Figure 8.1). It is readily recognizable by its dark green spiny foliage growing up to a metre in height. It is found in all areas, even at altitudes of 750 m, and on saline soil. The species is **dioecious**, the male producing spherical and the female slightly elongated purple flower heads from July to September. Only when the two sexes of plants are within about 100 m of each other does fertilization occur in sufficient quantities to produce large numbers of brown, shiny fruit, 4 mm long. These are wind-borne by a parachute of long hairs. The seeds may germinate beneath the soil surface in the same year as their production, or in the following spring, particularly when soil temperatures reach 20°C. The resulting seedlings develop a **tap root** which commonly reaches 3 m down into the soil. **Lateral roots** growing out horizontally about 0.3 m below the soil surface may penetrate 6 m in one season, and along their length are produced **adventitious buds** which, each spring, grow up as stems. Under permanent grassland, the roots may remain dormant for many years. Soil disturbance, such as ploughing, breaks up the roots and may result in a worse thistle problem unless drought or frost coincides with this activity.

Control of the seedling stage of this weed is not normally considered worthwhile. The main strategy must be to prevent the development of the perennial root system. Cutting down plants at the flower bud stage when sugars are being transferred from the roots upwards is said to achieve this objective. **Translocated** chemicals, e.g. MCPA, are commonly used on grassland, the spray chemical moving from the leaves to the perennial roots, particularly when applied in autumn at a time when sugars are similarly moving down the plant. In broad-leaved perennial crops, e.g. raspberries, **soil acting** herbicides, e.g. bromacil, may be applied in late winter before crop bud burst, to suppress the emergence of established plants. Deep ploughing is often successful in newly cultivated land.

**Couch grass** (*Agropyron repens*)

This grass, sometimes called 'twitch', is a widely distributed and important weed found at altitudes up to 500 m.

The dull-green plant is often confused, in the vegetative stage, with the creeping bent (*Agrostis stolonifera*). However, the small 'ears' at the leaf base characterize couch. The plant may reach a metre in height and often grows in tufts. Flowering heads produced from May to October resemble **perennial ryegrass**, but the flat flower spikelets are positioned at right angles to the main stem in couch. Seeds (9 mm long) are produced only after cross-fertilization between different strains of the species, and the importance of the seed stage, therefore, varies from field to field. Couch seeds may be carried in cereal seed stocks over long distances, and may survive deep in the soil for up to ten years. From May to October, stimulated by high light intensity, the overwintered plants produce horizontal **rhizomes** (*see* Figure 6.5) just under the soil; these white rhizomes may grow 16 cm per year in heavy soils, 32 cm in sandy soils. They bear scale leaves on nodes which, under **apical dominance** remain suppressed until the rhizome is cut by ploughing. In the autumn, rhizomes attached to the mother plant grow above ground to produce new plants which survive the winter. The rapid growth and extensive root system of this species provides severe competition for light, water and nutrients in any infested crop.

Control is achieved by a combination of methods. In fallow soil **deep ploughing**, especially in heavy land, exposes the rhizomes to drying. Further control by **rotavating** the weed when it reaches the one or two leaf stage disturbs the plant at its weakest point, and repeated rotavating will eventually cut up couch rhizomes to such an extent that no further nodes are available for its propagation.

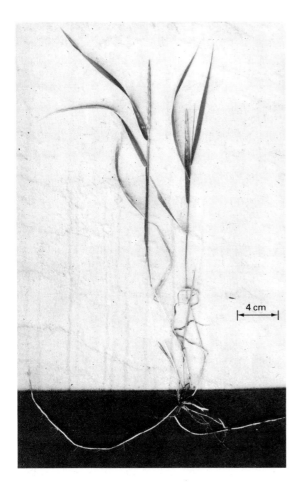

**Figure 8.5** *Couch grass: note the underground rhizome.*

**Translocated** herbicides, e.g. glyphosate sprayed on to couch in fallow soils during active vegetative growth, kill most of the underground rhizomes. In established fruit and conifers, control is achieved in the dormant winter or early spring seasons by a translocated chemical, e.g. dalapon, which, if applied to a full weed cover, has a maximum effect without causing root damage to the trees. Soil intended for vegetable production, e.g. carrots and potatoes, may be sprayed with a chemical such as dalapon to control couch so long as a seven-week period is observed between application and planting. Control by **contact** herbicides,

e.g. paraquat, may be successful in exhausting couch rhizome reserves if applied repeatedly.

**Yarrow** (*Achillea millifolium*)

This strongly scented perennial, with its spreading flowering head (Figure 8.6) is a common hedgerow plant found on most soils at altitudes up to 1200 m. Its persistence, together with its resistance to herbicides and drought in grassland, make it a serious turf weed.

The seedling leaves are hairy, and elongated with sharp teeth (Figure 8.1). The mature plant has dissected pinnate leaves produced throughout the year on wiry, woolly stems, which commonly reach 45 cm in height, and which from May to September produce flat-topped white to pink flower heads. An annual production of about 3000 small, flat fruits per plant is dispersed by birds. The seeds germinate on arrival at the soil surface. When not in flower, this species produces stolons along the ground (up to 20 cm long per year), and in autumn rooting from the nodes occurs.

Control of this weed may prove difficult. Routine scarification of turf does not easily remove the roots. Repeated sprays of **translocated** chemical mixtures, e.g. 2,4,-D plus mecoprop, reach the roots to give some control.

**Figure 8.6** *Yarrow: note the divided leaves and branching inflorescence.*

**Broad-leaved dock** (*Rumex obtusifolius*)

The seedling cotyledons are narrow, the true leaves large, broad and crimson-coloured (Figure 8.1). This is a common weed of arable land, grassland and fallow soil. The mature plant is readily identified by its long (up to 25 cm) shiny green leaves (Figure 8.7), known to many as an antidote to 'nettle rash'. The plant may grow 1 m tall, producing a conspicuous branched inflorescence of small green flowers from June to October. The numerous plate-like fruits (3 mm long) may fall to the ground or be dispersed by seed-eating birds such as finches, by cattle, and in

**Figure 8.7** *Broad-leaved dock: note the large, shiny leaves.*

batches of seed stocks. The seed represents an important stage in this perennial weed's life cycle, surviving many years in the soil, and most commonly germinating in spring. Like most *Rumex* spp., the seedling develops a stout, branched tap root, which may penetrate the soil down to 1 m in the mature plant, but most commonly reaches 25 cm. Segments of the tap root, chopped by cultivation implements, are capable of producing new plants.

High levels of seed production, a tough tap root, and a resistance to most herbicides, thus present a problem in the control of this weed. Attempts to exhaust the root system by repeated ploughing and rotavating have proved useful. Young seedlings are easily controlled by **translocated** chemicals, e.g. 2,4-D, but the mature plant is resistant to all but a few **translocated** chemicals, e.g. asulam, which may be used on grassland, soft fruit, top fruit, and amenity areas, during periods of active vegetative weed growth when the chemical is moved most rapidly towards the roots.

**Mixed weed populations**

In the field, a wide variety of both annual and perennial weeds may occur together. The grower must recognize the most important weeds in his holding or garden, so that a decision on the **precise** use of chemical control with the correct herbicide is achieved. Particular care is required to match the **concentration** of the herbicide or herbicide mixtures to the weed species present. Also, the grower must be aware that continued use of one chemical may induce a change in weed species, some of which may be tolerant to that chemical.

## MOSSES AND LIVERWORTS

These primitive plants may become weeds in wet growing conditions. The small cushion-forming moss (*Bryum* spp.) grows on **sand capillary** benches, and thin, acid turf which has been closely mown. Feathery moss (*Hypnum* spp.) is common on less closely mown, unscarified turf. A third type (*Polytrichum* spp.), erect and with a rosette of leaves, is found in dry acid conditions around golf greens. Liverworts (*Pellia* spp.) are recognized by their flat (thallus) leaves growing on the surface of pot plant compost.

These organisms increase only when the soil and compost surface is excessively wet, or when nutrients are so low as to limit plant growth. **Cultural** methods such as improved drainage, aeration, liming, application of fertilizer and removal of shade usually achieve good results in turf. Control with **contact** scorching chemicals, e.g. alkaline ferrous sulphate, may give temporary results. Moss on sand benches becomes less of a problem if the sand is regularly washed, although contact herbicides, e.g. chloroxuron, are available. Liverworts may be controlled in nursery stock by the herbicide oryzalin.

## PRACTICAL EXERCISES

1 Using a suitable identification book for reference, determine which are the most common weed species present as seedlings in a chosen horticultural area. March and April are suitable times for this exercise.

Compare the seedling flora in different horticultural situations.

Using suitable reference books and trade leaflets, determine the most effective cultural and chemical control to be used against the weeds that have been identified.

2 Examine a crop infested by weeds. Compare the size of plants in infested and non-infested areas. Weigh the weeds, and the crop plants, in a small area to obtain an estimate of plant loss.

3 Follow the growth of weed seedlings in pots to determine the time taken to produce seed, and the number of seeds produced by each plant.

Investigate the effects of mulches on weed development.

4 Carefully dig up perennial weeds and observe the size and detail of their perennating organs.

Chop up these organs to investigate the smallest piece able to survive in soil.

Investigate the depth of soil at which these organs can survive.

## FURTHER READING

British Crop Protection Council. *The UK Pesticide Guide* (1993).

Fryer, J.D., and Makepeace, R.J. *Weed Control Handbook*, Vol. 2 *Recommendations* (Blackwell Scientific Publications, 1978).

Hance, R.J., and Holly, K. *Weed Control Handbook*, Vol. 1 *Principles* (Blackwell Scientific Publications, 1990).

Hill, A.H. *The Biology of Weeds* (Edward Arnold, 1977).

Hope, F. *Turf Culture: A Manual for Groundsmen* (Cassell, 1990).

*Schering Guide to Grass and Broad-leaved Weed Identification* (1990).

Spiller, Mary. *Weeds – Search and Destroy* (MacDonald, 1985).

*Welcome to the World of Environmental Products* (Rhone-Poulenc, 1992).

# 9

# Large Pests

In this chapter the rabbit, rat, mole, deer, woodpigeon and bullfinch are described. Their size and mobility when compared with smaller pests (see Chapter 10) necessitates control against the individual pest organism, and the horticulturist must particularly avoid harbouring these pests which can move to neighbouring holdings. With the increased setting aside of cultivated land, an increase in large pest numbers can be expected.

## THE RABBIT (*Oryctolagus cuniculus*)

The rabbit is common in most countries of central and southern Europe. It came to Britain around the 11th century with the Normans, and became an established pest in the 19th century.

The rabbit may consume 0.5 kg of plant food per day. Young turf and cereal crops are the worst affected, particularly winter varieties which, in the seedling stage, may be almost completely destroyed. Rabbits may move from cereal crops to horticultural holdings. Stems of 'top' fruit may be **ring-barked** by rabbits, particularly in early spring when other food is scarce. Vegetables and recently planted garden-border plants are a common target for the pest, and fine turf on golf courses may be damaged, thus allowing lawn weeds, e.g. yarrow to become established.

The rabbit's high reproductive ability enables it to maintain high populations even when continued control methods are in operation. The doe, weighing about 1 kg, can reproduce within a year of its birth, and may have three to five litters of 3–6 young in one year, commonly in the months of February to July. The young are blind and naked at birth, but emerge from the underground 'maternal' nest after only a few weeks to find their own food. Large burrow systems (**warrens**), penetrating as deep as 3 metres in sandy soils, may contain as many as a hundred rabbits. Escape or **bolt** holes running off from the main burrow system allow the rabbit to escape from predators.

Control of rabbits is, by law, the responsibility of the land owner. **Preventative** measures are the most effective. **Wire fencing**, with the base 30 cm **underground** and facing outwards, represents an effective barrier to the pest, while thick plastic sheet **guards** are commonly coiled round the base of exposed young trees. Repellant chemicals, e.g. **colophene**, dissolved in alcohol, may be sprayed on to young trees or, when impregnated in sacking rolls, placed on short poles, to help protect ornamental bedding displays.

**Shotguns**, **small spring traps** placed in the rabbit hole, winter ferreting, or **long nets** placed at the corner of a field to catch herded rabbits, are methods used as **curative** control. **Gassing**, however, is the most effective method. Crystals of powdered cyanide are introduced by trained operators into the holes of warrens by means of long-handled spoons or by power operated machines. On contact with moisture, hydrocyanic acid is released as a gas and, in well-blocked warrens, the rabbits are quickly killed. Care is required in the storage and use of powdered cyanide, where an **antidote**, amyl nitrite, should be readily available. **Myxamatosis**, a fleaborne virus disease of the rabbit, causing a swollen head and eyes, was introduced into Britain in 1953, and within a few years greatly reduced the rabbit population. The development of weaker virus strains, and the increase in rabbit resistance, have combined to reduce this disease's effectiveness in control, although its importance in any one area is constantly fluctuating.

## DEER

Deer may become pests in land adjoining woodland where they hide. Muntjac and roe deer ring-bark trees and eat succulent crops. High fences and regular shooting may be used in their control.

## THE BROWN RAT (*Rattus norvegicus*)

The brown rat, also called the common rat, is well known by its dark-brown colour, blunt nose, short ears and long, scaly tail. Its diet is varied; it will eat **seeds**, **succulent** stems, bulbs and tubers, and may grind its teeth down to size by the unlikely act of gnawing at plastic **piping** and electric cables. A rat's average annual food intake may reach 50 kg, a large amount for an animal weighing only about 300 g. This species has considerable reproductive powers. The female may begin to breed at 8 weeks of age, producing an average of six litters of six young per year. Its unpopular image is further increased by its habit of fouling the food it eats, and by the lethal human bacterium causing **Weil's disease**, which it transmits through its urine.

Control is best achieved by a preliminary survey of rat numbers in buildings and fields of the horticultural holding, and by the identification of the 'rat-runs' along which the animals travel. Baits containing a mixture of **anticoagulant** poison and food material such as oatmeal are placed near the runs, inside a container which, while attracting the rat, prevents access by children and pets, Drainage tiles or oil drums drilled with a small rat-sized hole often serve this purpose. The poison, e.g. difenacoum, takes about three days to kill the rat and, since the other individuals do not associate their comrade's death with the chemical, the whole family may be controlled. The bait should be placed wherever there are signs of rat activity, and repeated applications every three days for a period of three weeks should be effective.

Strains of rat resistant to some anticoagulants are commonly found in some areas, and a range of chemicals may need to be tried before successful control is achieved. The poison, when not in use, should be safely stored away from children and pets. Dead rats should be burnt to avoid poisoning of other animals.

**Sonic** devices are sometimes used to disturb the animal and provide a round-the-clock deterrent.

## THE GREY SQUIRREL (*Sciurus carolinensis*)

This attractive-looking, 45 cm-long creature was introduced into Britain in the late 19th century, at a time when the red squirrel population was suffering from disease. The grey squirrel became dominant in most areas, with the red squirrel in pockets such as the Isle of Wight.

The horticultural damage caused by grey squirrels varies with each season. In spring, germinating bulbs may be eaten, and the **bark** of many tree species stripped off. In summer, pears, plums and peas may suffer. Autumn provides a large wild food source, though apples and potatoes may be damaged. In winter, little damage is done. Fields next to wooded areas are clearly prone to squirrel damage.

Squirrels most commonly produce two litters of three young from March to June, in twig platforms high in the trees; the female may become pregnant at an early age (six months). Since the squirrels have few natural enemies, and this species lives high above ground, control is difficult.

During the months April to July, when most damage is seen, **cage traps** containing desirable food, e.g. maize seed, reduce the squirrel population to less damaging levels. **Spring traps** placed in natural or artificial tunnels achieve rapid results at this time of year if placed where the squirrel moves. **Poisoned bait** containing 0.02 per cent anticoagulant chemicals, e.g. warfarin, when placed in a well-designed ground-level hopper (one hopper per 3 hectares) may achieve successful squirrel control without seriously lowering other small wild mammal numbers. In winter and early spring, the destruction of squirrel nests (or **dreys**) by means of long poles may achieve some success.

## THE MOLE (*Talpa europea*)

The mole is found in all parts of Great Britain except Ireland. This dark-grey, 15 cm-long mammal, weighing about 90 g, uses its shovel-shaped feet to create an underground system 5–20 cm deep and up to 0.25 h in extent. The tunnel contents are excavated into mole hills. The resulting **root disturbance** to grassland and other crops causes wilting, and may result in serious losses.

In its dark environment, the solitary mole moves, actively searching for earthworms, slugs, millipedes and insects. About five hours of activity is followed by about three hours of rest. Only in spring do males and females meet. In June, one litter of two to seven young are born in a grass-lined underground nest, often located underneath a dense thicket. Young moles often move above ground, reach maturity at about four months, and live for about four years.

Natural **predators** of the mole include tawny owls, weasels and foxes. The main control methods are **trapping** and **poison baiting**, usually carried out between October and April, when tunnelling is closer to the surface. Pincer or half barrel traps are placed in fresh tunnels and sprung without greatly changing the tunnel diameter. The soil must be replaced so that the mole sees no light from its position in the tunnel. The mole enters the trap, is caught and starves to death. In serious mole infestations, **strychnine salts** are mixed with earthworms at the rate of 2 g per 100 worms, and single worms carefully inserted into inhabited tunnels at the rate of 25 worms per hectare. Ministry of Agriculture authority is required before purchasing strychnine, a highly dangerous chemical, which must be stored with care.

## THE WOOD-PIGEON (*Columba palumbus*)

This attractive, 40 cm-long, blue-grey pigeon with white underwing bars is known to horticulturists as a serious pest on most outdoor edible crops. In **spring**, seeds and seedlings of crops such as brassicas, beans and germinating turf may be systematically eaten. In **summer**, cereals and clover receive its attention; in **autumn**, tree fruits may be taken in large quantities, while in **winter**, cereals and brassicas are often seriously attacked, the latter when snowfall prevents the consumption of other food. The wood-pigeon is invariably attracted to **high protein** foods such as seeds when they are available.

Wood-pigeons lay several clutches of two eggs per year from March to September. The August/September clutches show highest survival. The eggs, laid on a nest of twigs situated deep inside the tree, hatch after about 18 days, and the young remain in the nest for 20–30 days. Predators such as jays and magpies eat many eggs, but the main population control factor is the availability of food in winter. Numbers in Great Britain are boosted a little by migrating Scandinavian pigeons in April, but the large majority of this species are resident and non-migratory.

The wood-pigeon spends much of its time feeding on wild plants, and only a small proportion of its time on crops. Control of the whole population, therefore, seems both costly and impracticable. Protection of particular fields is achieved by **scaring** devices which include scarecrows, bangers (firecrackers or gas guns), or rotating orange and black vanes, which disturb the pigeon. Changing the **type** and **location** of the device every few days helps prevent the pigeon from becoming indifferent. The use of the **shotgun** from hidden positions such as hides and ditches, particularly when artificial pigeons (decoys) are placed in the field, is an important additional method of scaring birds and thus protecting the crop.

## THE BULLFINCH (*Pyrrhula pyrrhula*)

This delightful, 14 cm-long bird is characterized by its sturdy appearance and broad bill. The male has a rose-red breast, blue-grey back, and black headcap. The female has a less striking pink breast, and yellowish-brown back. From April to September the bird progressively feeds on seeds of wild plants, e.g. chickweed, buttercup, dock, fat hen and blackberry. From September to April, the species forms small flocks which, in addition to feeding on buds and seeds of wild species, e.g. docks, willow, oak and hawthorn, turn their attention to **buds** of soft and top fruit. Gooseberries are attacked from November to January, apples from February to April, and black-currants from March to April. The birds are shy, preferring to forage on the edges of orchards but, as winter advances, become bolder, moving towards the more central trees and bushes. The birds nip buds out at the rate of about 30 per minute, eating the central meristem tissues. Leaf, flower and fruit development may thus be seriously reduced, and since in some plums and gooseberries there is no regeneration of fruiting points, damage may be seen several years after attack.

The bullfinch produces a platform nest of twigs in birch or hazel trees, and between May and September lays two to three clutches of 4−5 pale blue eggs, with purple-brown streaks, and can thus quickly re-establish numbers reduced by lack of food or human attempts at control.

Fine mesh netting, cotton or synthetic thread draped over trees, or bitter chemicals, e.g. thiram, sprayed at the time of expected attack, are used to some extent to prevent bullfinch attack. Most success is achieved by catching birds (usually immature individuals) in specially designed **traps**, which are closed when the bird lands on a perch to eat seeds. Trapping may be started as early as September. Large-scale re-invasion by the birds is unlikely, as they are territorial, rarely moving more than 2 miles throughout their lives; thus large numbers may be killed before damage begins to occur. Bullfinch control is permitted only in scheduled areas of wide-scale fruit production, e.g. Kent.

## PRACTICAL EXERCISES

1 Search for a rabbit warren. Look for damage in neighbouring crops.
2 Observe the effectiveness of wood pigeon scaring devices.
3 Look for bullfinches in an orchard. Note the damage they cause.

## FURTHER READING

MAFF Advisory Leaflets:
   No. 165 *The Woodpigeon* (1978)
   No. 234 *The Bullfinch* (1973)
   No. 318 *The Mole* (1975)
   No. 534 *The Rabbit* (1988)
   No. 608 *Control of Rats on Farms* (1987)
Murton, R.K. *The Woodpigeon* (Collins, 1982).
Rowe, J.J. *Grey Squirrel Control* (Forestry Commission, 1973).
Thompson, H.V., and Worden, A.N. *The Rabbit* (Collins, 1956).
Wright, E.N., Inglis, I.R., and Feare, C.J. *Bird Problems in Agriculture* (British Crop Protection Council, 1980).

# 10

# Small Pests, Soil Organisms and Bees

*Slugs, insects, mites, related small pests and nematodes may do serious damage to horticultural crops. This chapter describes damage, life cycle details and specific control measures relevant to the thirty-three important species. A more general description of control measures is given in Chapter 12. Mention is made of beneficial animals occurring in the soil, and bees, useful in pollination.*

## SLUGS

These animals belong to the phylum Mollusca, a group including the octopus and whelk, and the slug's close relatives, the **snails**, which cause a little damage to plants in greenhouses and private gardens. Slugs move slowly by means of an undulating foot. Unlike the snail, their lack of shell permits movement through the soil in search of their food source, seedlings, roots, tubers and bulbs. The slug feeds by means of a file-like tongue (**radula**) which cuts through plant tissue held by the soft mouth, and scoops out cavities in the affected plants (Figure 10.1). In moist, warm weather it may cause above-ground damage to leaves of plants such as establishing turf, lettuce and Brussels sprouts. Slugs are **hermaphrodite** (bearing in their bodies both male and female organs), mate in spring and summer, and lay clusters of up to fifty round, white eggs in rotting vegetation, the warmth from which protects this sensitive stage during cold periods.

Slugs range in size from the black keeled slug (*Milax*), 3 cm long, to the garden slug (*Arion*) which reaches 10 cm in length. Mottled carnivorous slugs (*Testacella*) are occasionally found feeding on earthworms. Horticultural areas commonly support populations of 50 000 slugs per hectare.

Many non-chemical forms of control have been used, ranging from baits of grapefruit skins and stale beer to soot sprinkled around larger plants. The most effective methods, however, involve the two chemicals **metaldehyde**, which dries the slug out, and **methiocarb**, which acts as a stomach poison. The chemicals are most commonly used as small coloured **pellets** (which include attractants such as bran and sugar), but metaldehyde may also be applied as a drench. Some growers estimate the slug population using small heaps of pellets covered with a tile or flat stone (to prevent bird poisoning) before deciding on general control.

## INSECTS

**Belonging to the large group Arthropoda, which include also the woodlice, mites, millipedes (Table 10.1) and symphilids, the insects are horticulturally the most important group, particularly as pests,**

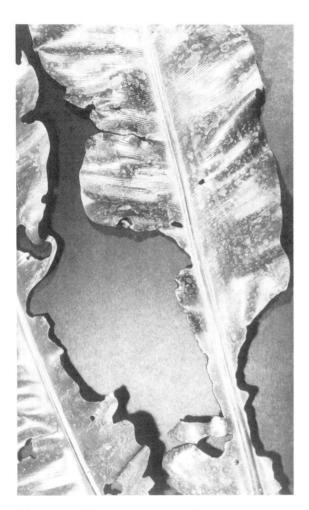

**Figure 10.1** *Slug damage on pot plant.*

**but also as beneficial soil animals. The body of the adult insect is made up of segments, and is divided into three main parts, the head, thorax and abdomen (Figure 10.2).**

### Structure and biology

The **head** bears three pairs of moving mouthparts. The first, the **mandibles** in insects such as caterpillars and beetles, have a biting action (Figure 10.3). The second and third pairs, the **maxillae** and **labia** in these insects, help in pushing food into the mouth. In the aphids, the mandibles and maxillae form a delicate tubular stylet, which sucks up liquids from the plant phloem tissues. Insects remain aware of their environment by means of compound eyes, and antennae which have a touching and smelling function.

The **thorax** bears three pairs of legs, and in most insects two pairs of wings.

The **abdomen** bears lateral breathing holes (spiracles), which lead to a respiratory system of tracheae. The blood is colourless, circulates digested food and has no respiratory function. The digestive system, in addition to its food absorbing role, removes waste cell products from the body by means of fine, hair-like growths (Malpighian tubules) located near the end of the gut.

An insect may develop from egg to adult in one of two ways. Since the animal has an **external** skeleton made of tough chitin, it must shed and replace its skin (**cuticle**, Figure 10.2) periodically by a process called **ecdysis**, in order to increase in size. Two to seven growth stages (**instars**) occur before the adult emerges. In the first group, the **aphids**, the egg hatches to form a first instar called a **nymph** (Figure 10.4), resembling the adult in all but size, wing development, and possession of sexual organs. Successive nymph instars more closely resemble the adult.

In the moths, butterflies, flies and beetles, however, the egg hatches to form a first instar, called a **larva** (Figure 10.4) which may differ greatly in shape from the adult, e.g. the larva (caterpillar) of the cabbage white bears little resemblance to the adult butterfly. The great change (**metamorphosis**) necessary to achieve this transformation occurs inside the **pupa** stage.

The method of overwintering differs between insect groups. The aphids survive mainly as the eggs, while most moths, butterflies and flies survive as the pupa (chrysalis).

The speed of increase of insects varies greatly between groups. Aphids may take as little as twenty days to complete a life cycle in summer, often resulting in vast numbers in the period June to September. On the other hand, the wireworm,

**Table 10.1**   Arthropod groups found in horticulture

| Group | Key features of group | Habitat | Damage |
|---|---|---|---|
| **Woodlice** (*Crustacea*) | Grey, 7 pairs of legs, up to 12 mm in length | Damp organic soils | Eat roots and lower leaves |
| **Millipedes** (*Diplopoda*) | Brown, many pairs of legs, slow moving | Most soils | Occasionally eat underground tubers and seed |
| **Centipedes** (*Chilopoda*) | Brown, many pairs of legs, very active, with strong jaws | Most soils | Beneficial |
| **Symphilids** (*Symphyla*) | White, 12 pairs of legs, up to 8 mm in length | Glasshouse soils | Eat fine roots |
| **Mites** (*Acarina*) | Variable colour, usually have 4 pairs of legs (e.g. red spider mites) | Soils and plant tissues | Mottle or distort leaves, buds, flowers and bulbs. Soil species are beneficial |
| **Insects** (*Insecta*) | Usually 6 pairs of legs, 2 pairs of wings | | |
| Springtails (*Collembola*) | White to brown, 3–10 mm in length | Soils and decaying humus | Eat fine roots. Some beneficial |
| Aphid group (*Hemiptera*) | Variable colour, sucking mouthparts, produce honeydew (e.g. greenfly) | All habitats | Discolour leaves and stems. Prevent flower pollination. Transmit viruses |
| Moths and Butterflies (*Lepidoptera*) | Large wings. Larva with 3 pairs of legs, and 4 pairs of false legs and biting mouthparts (e.g. cabbage white butterfly) | Mainly leaves and flowers | Defoliate leaves (stems and roots) |
| Flies (*Diptera*) | One pair of wings, larvae legless (e.g. leatherjacket) | All habitats | Leaf mining, eat roots |
| Beetles (*Coleoptera*) | Horny front pair of wings, well-developed mouthparts in adult and larva (e.g. wireworm) | Mainly in the soil | Eat roots and succulent tubers (and fruit) |
| Sawflies (*Hymenoptera*) | Adult like a Queen Ant. Larvae have 3 parts of legs, and more than 4 pairs of false legs (e.g. rose-leaf curling sawfly) | Mainly leaves and flowers | Defoliation |
| Thrips (*Thysanoptera*) | Yellow and brown, very small, wriggle their bodies (e.g. onion thrips) | Leaves and flowers | Cause spotting of leaves and petals |
| Earwigs (*Dermaptera*) | Brown, with pincers at rear of body | Flowers and soil | Eat flowers |

the larva of the click beetle, taking four years to complete its life cycle, increases much more slowly.

Insect groups are classified into their appropriate order (Table 10.1) according to their general appearance and life cycle stages. A selection of insect pests is presented below, each species having particular features of its life cycle which warrant description. Some damaging larval stages are compared in Figure 10.5. Whilst comments on **control** are mentioned, the reader should refer to Chapter 12 for details of specific types of control (cultivations, chemicals etc.) and for explanations of terms used.

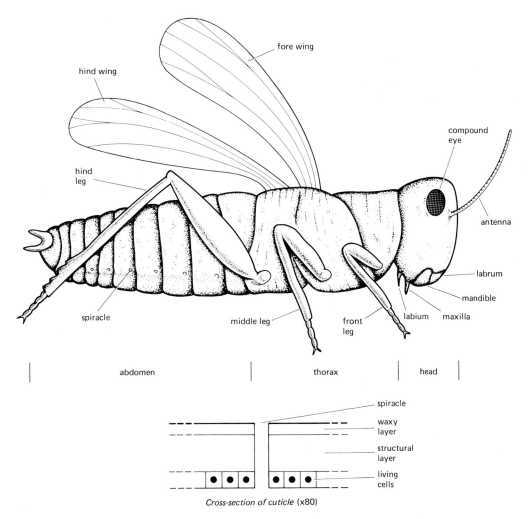

**Figure 10.2** *External appearance of an insect. Note the mouthparts, spiracles and cuticle, the three main entry points for insecticides.*

**Springtails (order Collembola)**

This group of primitive wingless insects, about 2 mm in length (Figure 10.24), have a springlike appendage at the base of the abdomen. They are very common in soils, and normally aid in the breakdown of **soil organic** matter. Two genera, *Bourletiella* and *Collembola*, however, may do serious damage to conifer seedlings, and cucumber roots respectively. With a serious infestation, a residual chemical, e.g. HCH, is incorporated into compost or drenched into the soil.

**Aphids and their relatives (order Hemiptera)**

This important group of insects has the egg-nymph-adult life cycle, and sucking mouth-parts.

**Peach-potato aphid** (*Myzus persicae*, often referred to by the name 'greenfly') is common in

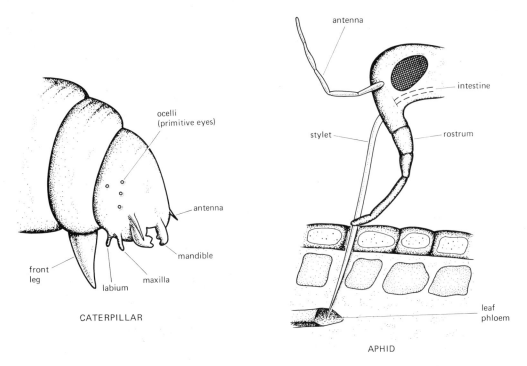

**Figure 10.3** *Mouthparts of the caterpillar and aphid. Note the different methods of obtaining nutrients. The caterpillar consumes all the plant material it meets. The aphid selectively sucks up dilute sugar solution from the phloem tissues.*

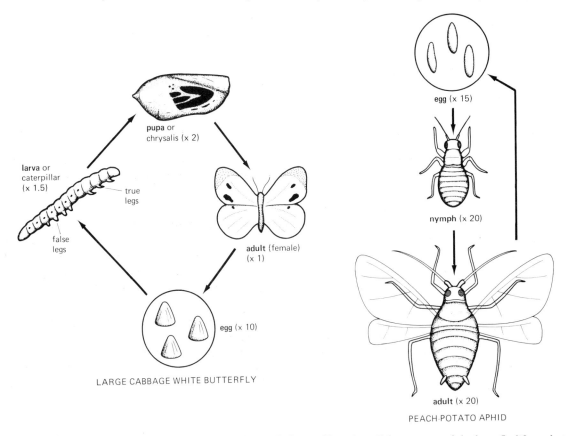

**Figure 10.4** *Life cycle stages of a butterfly and an aphid pest. Note that all four stages of the buttefly life cycle are very different in appearance. The nymph and adult of the aphid are similar.*

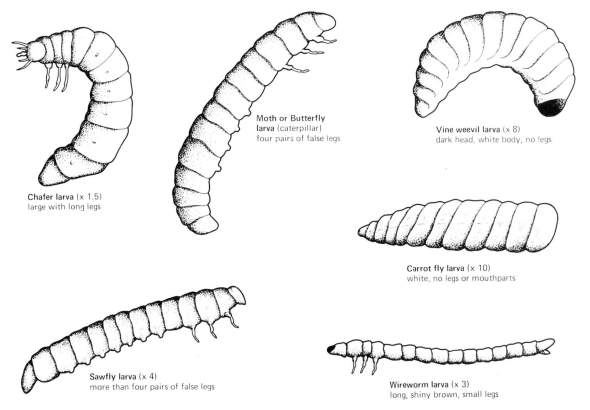

**Figure 10.5** *Insect larva which damage crops. Identification into the groups above can be achieved by observing the features of colour, shape, legs and mouthparts.*

market gardens and greenhouses. It varies in colour from light green to orange, measures 3 mm in length (Figure 10.6), and has a complex life cycle, shown in Figure 10.7, alternating between the winter host (peach) and the summer hosts, e.g. potato and bedding plants. In spring and summer, the females produce nymphs directly without any egg stage (a process called **vivipary**), and without fertilization by a male (a process called **parthenogenesis**). Only in autumn, in response to decreasing daylight length and outdoor temperatures, are both sexes produced which, having wings, fly to the winter host, the peach. After the female is fertilized, she lays thick-walled eggs. In glasshouses, the aphid may survive the winter as the nymph and adult female on plants such as begonias and chrysanthemums, or on weeds such as fat hen.

The nymph and adult of this aphid may cause three types of damage. Using its sucking stylet, it may inject a digestive juice into the plant phloem, which in young organs may cause severe **distortion**. Having sucked up sugary phloem contents, the aphid excretes a sticky substance called **honeydew**, which may block up leaf stomata and reduce photosynthesis, particularly when dark-coloured fungi (**sooty moulds**) grow over the honeydew. Thirdly, the aphid stylets may transmit viruses such as virus Y on potatoes, and tomato aspermy virus on chrysanthemums.

Peach-potato aphid is controlled in three main ways. In outdoor crops, several organisms, e.g. ladybirds, lacewings, hoverflies, and parasitic fungi (*see* **Biological control**, Chapter 12), naturally found in the environment, may reduce the pest's importance in favourable seasons. In the glasshouse, a parasitic wasp, *Aphidius matricariae*, commonly controls the aphid. **Contact** chemicals, e.g. malathion, are able to penetrate the thin cuticle of the insect, while other chemicals, e.g. dimethoate, moving through the internal tissues of the plant (called **systemic** chemicals) are sucked up by the aphid stylet and reach the insect's digestive system. **Fumigant** chemicals, e.g. nicotine, enter the insect through the spiracles.

*Spruce-larch adelgid* (*Adelges viridis*). This relative of the aphid may cause serious damage on spruce grown for Christmas trees.

The green adult develops from overwintering nymphs on **spruce**, and in May (year one) lays about fifty eggs on the dwarf shoots. The emerging nymphs, injecting poisons into the shoots, cause abnormal growth into green **pineapple galls**, which spoil the tree's appearance. Although nursery trees of less than four years of age are rarely badly damaged, early infestation in the young plant may result in serious damage as it gets older. In June to September the adults move to **larch**, acquire woolly white hairs and may cause defoliation of

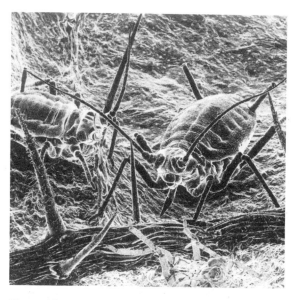

**Figure 10.6** *Peach-potato aphid: note the sucking stylet (Courtesy Mrs P. Evans, Rothamsted Experimental station.)*

the leaves. After a further year (year two) on this host, the adelges returns to the **spruce**, where it lives for another year (year three) before the gall-inducing stages are produced.

In Christmas trees, the adelges may be controlled by sprays of HCH in May, when the gall-inducing nymphs are developing.

There are many other horticulturally important aphid species. The black bean aphid (*Aphis fabae*), which overwinters on *Euonymus* bushes, may seriously damage broad beans, runner beans and red beet. The rose aphid (*Macrosiphum rosae*) attacks young shoots of rose.

*Glasshouse whitefly* (*Trialeurodes vaporariorum*). This small, moth-like pest was originally introduced from the tropics, and now causes serious problems on a range of glasshouse food and flower crops. It should not be confused with the slightly larger cabbage whitefly, a less important pest on brassicas.

The adult glasshouse whitefly (Figure 10.8) is about 1 mm long. The fertilized female lays about 200 minute, white, elongated oval eggs in a circular pattern on the lower leaf surface. After turning black, the eggs hatch to produce nymphs (**crawlers**), which soon become flat immobile **scales**, the last instar being a thick walled 'pupa' from which the adult emerges, and three days later lays eggs again. All stages after the egg have sucking stylets, which may cause large amounts of honeydew and sooty moulds on the leaf surface. The whole life cycle takes thirty-two days in spring, and twenty-three days in the summer.

Plants which are seriously attacked include fuchsias, cucumbers, chrysanthemums and pelargoniums. Chickweed, a common greenhouse weed, may harbour the pest overwinter in all stages of the life cycle.

Control of glasshouse whitefly is achieved in two main ways. A minute wasp (*Encarsia formosa*) lays an egg inside the scale of the whitefly which, being eaten away internally, turns black to release a wasp. **Chemical** control methods include soil-applied **systemic granular** compounds, e.g. oxamyl, where the chemical enters the whitefly stylet via the plant phloem. **Fumigant** chemicals,

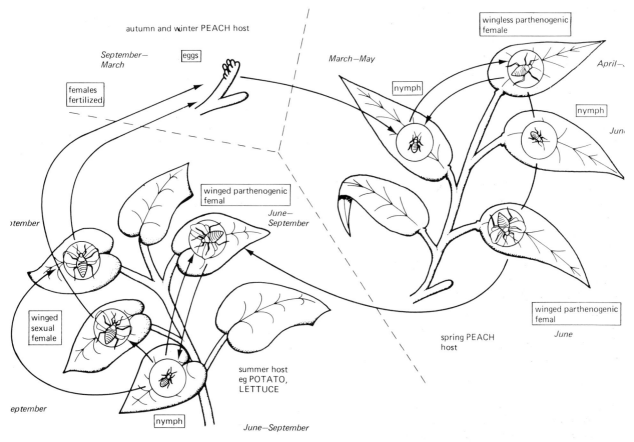

autumn and winter PEACH host

September–March

eggs

females fertilized

March–May

nymph

wingless parthenogenic female

April–

nymph

Jun

winged parthenogenic femal

June–September

)tember

winged sexual female

summer host eg POTATO, LETTUCE

winged parthenogenic femal

June

spring PEACH host

eptember

nymph

June–September

**Figure 10.7** *Peach-potato aphid life cycle throughout the year. Female aphids produce nymphs on both the peach and summer host. Winged females develop from June to September. Males are produced only in autumn. Eggs survive the winter. In glasshouses, the life cycle may continue throughout the year.*

e.g. resmethrin and pirimiphos-methyl, quickly kill the scale and adult stages, but the eggs and 'pupae' are little affected. It is therefore suggested that serious infestations of this pest receive a regular chemical spray or fog at three-day intervals for a month to control emergent crawlers and adults.

***Greenhouse mealy bug*** (*Planococcus citri*). This pest, a distant relative of the aphid, spoils the appearance of some glasshouse crops, particularly orchids, *Coleus* species, cacti and *Solanum* species. Being a tropical species, it develops most quickly in high temperatures and humidities, and at 30°C completes a life cycle within about twenty-

two days. The adult measures about 3 mm in length (Figure 10.9), and produces fine waxy threads. All of the stages except the egg suck phloem juices by means of a tubular mouthpart (**stylet**), and when this pest is present in dense masses it produces **honeydew** and causes leaf drop on most plant species. Mealy bugs are difficult pests to control, as the thick cuticle resists chemical sprays, and the droplets fall off the waxy threads. A **contact** chemical, e.g. diazinon, may be sprayed if the plant species is not damaged by the chemical. **Systemic** chemicals, e.g. aldicarb, may be applied to the compost and reach the pest through the plant vascular tissues and stylet of

**Figure 10.8** *Adult whitefly. (Courtesy Shell Chemicals.)*

the pest. An introduced tropical ladybird, *Cryptolaemus montrouzieri*, is effective in controlling the pest above 20°C.

**Brown scale** (*Parthenolecanium corni*). The female scale, measuring up to 6 mm, is tortoise-shaped (Figure 10.10), and has a very thick cuticle. On fruit trees, e.g. apple, it is rarely seen now because of regular insecticide sprays, but in private gardens it may be a serious pest on vines, currants and cotoneasters, and in greenhouses attacks peaches and *Amaryllis*, causing stunted growth and leaf defoliation. As with mealybug, control is difficult because of the thick cuticle.

**Figure 10.9** *Mealy bug on pot plant. (Courtesy Plant Pathology Laboratory, MAFF.)*

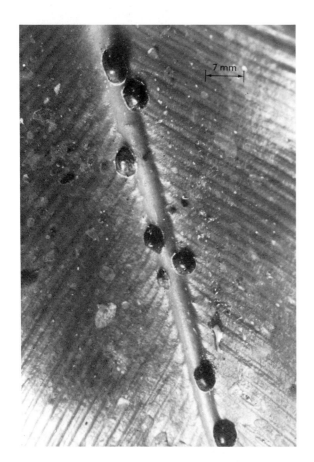

**Figure 10.10** *Brown scale on a pot plant.*

***Leaf hoppers*** (*Graphocephala fennahi*). These slender, light-green insects, about 3 mm long, well known in their nymph stage as 'cuckoo spit', are found on a wide variety of crops, e.g. potato, rose, *Primula* and *Calceolaria*. They live on the undersurface of leaves, causing a mottling of the upper surface. In strawberries, they are vectors of the green-petal disease, while in rhododendron they carry the serious bud blast disease which kills off the flower buds. August and September sprays of the residual chemical, HCH, prevent egg laying inside buds of rhododendron, and thus reduce the entry points for the fungus disease.

***Common green capsid*** (*Lygocoris pabulinus*). This active, light-green pest, measuring 5 mm in length, occurs on fruit trees and flower crops, most commonly outdoors. Because of the poisonous nature of its salivary juices, young foliage shows distorted growth with small holes, even when relatively low insect numbers are present (Figure 10.11), and fruit is scarred. This pest may be controlled by chemicals used against aphids.

**Thrips (order Thysanoptera)**

***Onion thrips*** (*Thrips tabaci*). This 1 mm-long, narrow-bodied insect has feather-like wings. Due to its great activity during warm, humid weather, it is sometimes called 'thunder fly'. Its mouthparts are modified for piercing and sucking, and the toxic salivary juices cause silvering in onion leaves, straw-brown spots on cucumber leaves (Figure

**Figure 10.11** *Capsid damage on rose: note the distorted foliage and holes in the leaves.*

**Figure 10.12** *Onion thrip damage on cucumber leaf. The distinct light-coloured spots differ from red spider mite damage (Figure 10.25).*

10.12), and white streaks in carnation blooms. The last instar of the life cycle, called the **pupa**, occurs in the soil, and it is this stage which over-winters. In greenhouses there may be seven generations per year, while outdoors one life cycle is common. The occurrence in Britain of Western flower thrip (*Frankliniella occidentalis*) on both greenhouse and outdoor flower and vegetable crops has created serious problems for the industry.

Thrips are sensitive to a wide range of insecticides, e.g. diclorvos, applied as a spray or fog. In greenhouse crops employing biological control, a drench of a **residual** insecticide, e.g. permethrin incorporated into a sticky solution, reduces insecticide contact with the predator or parasite.

## Earwigs (Forficula auricularia)

These pests belong to the order Dermaptera, and bear characteristic pincers (cerci) at the rear of the 15-mm long body (Figure 10.13). They gnaw away at leaves and petals of crops such as beans, beet, chrysanthemums and dahlias, usually from July to September, when the nymphs emerge from the parental underground nest. Upturned flower pots containing straw are sometimes used in glasshouses for trapping these shy nocturnal insects. Chemicals, e.g. HCH, able to penetrate the thick cuticle, are applied in the form of sprays or smokes, but flower scorch must be avoided.

## Moths and butterflies

This order (Lepidoptera) characteristically contains adults with four large wings and curled feed-

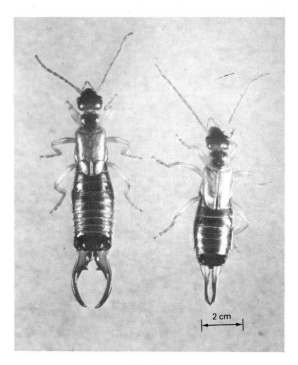

**Figure 10.13** *Larger earwig is male, smaller is female (Courtesy Shell Chemicals).*

ing tubes. The larva (**caterpillar**), with six small legs and eight false legs, is modified for a leaf-eating habit (Figure 10.4). Some species, however, are specialized for feeding inside fruit (codling moth on apple), underground (cutworms), inside leaves (oak leaf miner), or inside stems (leopard moth). The larva is the only damaging stage of this insect group.

*Large cabbage white butterfly* (*Pieris brassicae*). This well-known pest on cruciferous plants emerges from the overwintering pupa (chrysalis) in April and May and, after mating, the females lay batches of twenty to a hundred yellow eggs on the underside of leaves. Within a fortnight, groups of first instar larvae emerge and soon moult to produce the later instars, which are 25 mm long, yellow or green in colour, with clear black markings, and have well-developed mandibles. Pupation occurs usually in June, in a crevice or woody stem, the pupa (chrysalis) being held to its host by silk threads. A second generation of the adult commences in July, causing more damage than the first. The second pupa stage overwinters. The defoliating damage of the larva may result in skeletonized leaves of cabbage, cauliflower, Brussels sprouts and other hosts such as wallflowers and the shepherd's purse weed. Care should be taken not to confuse the cabbage white larva with the large smooth green or brown larva of the **cabbage moth,** or the smaller light green larva of the **diamond-backed moth,** both of which may enter the hearts of cabbages and cauliflowers, presenting greater problems for control.

There are several forms of control against the cabbage white butterfly. A small wasp (*Apantales glomeratus*) lays its eggs inside the pest larva (*see* parasites). A virus disease may infect the pest, causing the larva to go grey and die. Birds such as starlings eat the plump larvae. When damage becomes severe, the larvae may be controlled with sprays of insecticides, e.g. azinphos-methyl, or derris, containing extra **wetter/spreader,** to maintain the spray droplets on the waxy brassica leaves.

*Winter moths* (*Operophthera brumata*). This pest, which may be serious on top fruit and ornamental members of the Rosaceae family, emerges as the adult form from its soil-borne pupa in November and December. The male is a greyish brown moth, 2.5 cm across its wings, while the female is **wingless.** The female crawls up the tree to lay 100−200 light-green eggs in the bud clusters, which hatch at bud burst to produce green larvae with faint white stripes. These larvae, which move in a looping fashion (Figure 10.14), eat the leaf and form other leaves into loose webs, reducing the plant's photosynthesis. They occasionally scar young apple fruit before descending on a silk thread at the end of May to pupate in the soil until winter.

**Grease bands** wound around the main trunk of the tree in October prove very effective in preventing the female moth's progress. In large orchards, springtime sprays of an insecticide, e.g. carbaryl, kill the young instars of the larva.

**Figure 10.14** *Winter moth damage on apple: note the looping caterpillar of this species (Courtesy Plant Pathology Laboratory, MAFF).*

*Cutworms* (e.g. *Agrotis segetum*), the larvae of the **turnip moth**, unlike most other moth larvae, live in the soil, nipping off the stems of young plants and eating holes in succulent crops, e.g. bedding plants, lawns, potatoes, celery, turnips and conifer seedlings. The damage resembles that caused by slugs. The adult moth, 2 cm across, with brown forewings and white hindwings, emerges from the shiny soil-borne chestnut brown pupa in June to July, and lays about 1000 eggs on the stems of a wide variety of weeds. The first instar larvae, having fed on the weeds, descend to the soil and eventually reach 3.5 cm in length. They are grey to grey-brown in colour, with black spots along **the sides** (Figure 10.15). Several other cutworm **species**, e.g. heart and dart moth

(*Agrotis exclamationis*) and yellow underwing moth (*Noctua pronuba*) may cause damage similar to that of the turnip moth. This typically caterpillar-like larva should not be confused with the legless **leatherjacket** (Figure 10.19). There are normally two life cycles per year, but in hot summers this may increase to three.

Good weed control reduces cutworm damage. Soil drenches of **residual** insecticides, e.g. HCH; mixing of bran baits with stomach insecticides, e.g. fenitrothion; or broadcasting of methiocarb slug pellets, have all proved successful against the larva stage of this pest.

*Leopard moth* (*Zeuzera pyrina*) has an unusual life cycle. The adult female moth, 5–6 cm across, is white with black spots, and in early summer lays dark yellow eggs on the bark of apples, ash, birch, lilac and many other tree species. The emerging larva enters the stem by a bud, and then tunnels for two or three years in the **heartwood** before reaching 5 cm in length (Figure 10.16), pupating in the tunnel, and completing the life cycle as the adult. The tunnelling may weaken the branches of trees which, in high winds, commonly break. Where tunnels are observed, a piece of wire may be pushed along the tunnel to kill the larva, or a fumigant chemical, e.g. paradichlorbenzene crystals, may be placed in the tunnel before sealing it off with moist soil.

**Figure 10.15** *Cutworm larva of the turnip moth (Courtesy Plant Pathology Laboratory, MAFF).*

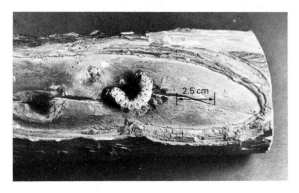

**Figure 10.16** *Leopard moth larva emerging from an apple stem.*

## Flies

This order (Diptera) is characterized by the single pair of clear forewings, the hind wings being adapted as balancing organs (halteres). The larvae are legless, elongated, and their mouthparts, where present, are simple hooks. The larvae cause crop damage.

*Carrot fly* (*Psila rosae*) is a widespread and serious pest on umbelliferous crops (carrots, celery, parsnips). The adult fly, shiny black with a red head, and 8 mm long, emerges from the soil-overwintering pupa in late May to early June. The small eggs laid on the soil near the host soon hatch to give white larvae (Figure 10.17), which eat fine roots and then enter the mature root, using fine hooks in their mouths. The mature larva, when a month old, leaves the host to turn into the cylindrical pale yellow pupa. A second

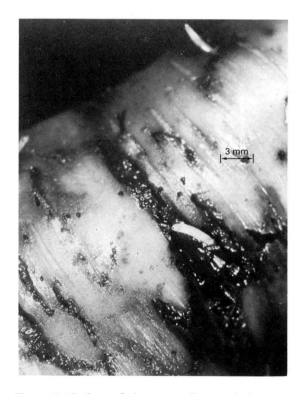

**Figure 10.17** *Carrot fly larva tunnelling inside the carrot root (Courtesy P.R. Ellis).*

generation of adults emerges in late July, while a third October emergence is seen in some areas.

Damage is similar in all crops. In carrots, seedlings may be killed, while in older plants the foliage may become red, and wilt in dry weather. Stunting is often seen, and affected roots, when lifted, are riddled with small tunnels which make the carrots unsaleable. Damage should not be confused with **cavity spot**, a condition associated with *Pythium* species of fungi, which produces elongated sunken spots around the root.

Damaging levels of carrot fly can be avoided by keeping hedges and nettle beds trimmed to reduce sheltering sites for the flies; by planting carrots after the May emergence has occurred; and by the incorporation of **residual** insecticides into the soil as granules, e.g. phorate, at planting time, or as a drench, e.g. chlorfenvinphos, after planting. Celery transplants may be dipped with a residual chemical, e.g. HCH.

*Chrysanthemum leaf miner* (*Phytomyza syngenesiae*). The leaf miners are a group of small flies, the larvae of which can do serious damage to horticultural crops. This species is found on members of the plant family Compositae, and the hosts attacked include chrysanthemum, cineraria and lettuce. Weed hosts, e.g. groundsel and sowthistle, may harbour the pest.

The flies emerge at any time of the year in greenhouses, but normally only between July and October outdoors. These adults, which measure about 2 mm in length and are grey-black with yellow underparts, fly around with short hopping movements. The female lays about 75 minute eggs singly inside the leaves, causing white spot symptoms to appear on the upper leaf surface. The **larva** stage is greenish white in colour, and tunnels into the palliside mesophyll of the leaf, leaving behind the characteristic mines seen in Figure 10.18. On reaching its final instar, the 3.5 mm long larva develops within the mine into a brown pupa, from which the adult emerges. The total life cycle period takes about three weeks during the summer months.

Certain chrysanthemum cultivars, e.g. Tuneful, may be badly attacked, while others, e.g. Yellow

**Figure 10.18** *Chrysanthemum leaf miner damage: note the light-coloured tunnels caused by the larva of this species (Courtesy Glasshouse Crops Research Institute).*

Iceberg, often resist the attack by the pest. While chemical sprays used against other pests, e.g. aphids and whitefly, may help in control of the adult, the larva stage is commonly controlled by a chemical, e.g. diazinon, which penetrates the leaf and reaches the active larva, or by **systemic** insecticide, e.g. aldicarb, which, after application as granules to the soil, then moves up to the leaf.

The occurrence of South American leaf miner (*Liriomyza huidobrensis*) and American serpentine leaf miner (*Liriomyza trifolii*) on a wide variety of greenhouse plants has created serious problems for the industry.

***Leatherjacket*** (*Tipula paludosa*) an underground pest, is a natural inhabitant of grassland and causes problems on fine turf. The ploughing up of grassland may, however, result in the pest damaging following crops such as potatoes, cabbages, lettuce and strawberries.

The adult of this species is the **crane fly**, or 'daddy-long-legs', commonly seen in August. The females lay up to 300 small eggs on the surface of the soil at this period, and the emerging larvae feeding on plant roots during the autumn, winter,

and spring months reach lengths of 4 cm by June. They are cylindrical, grey-brown in colour, legless, and possess hooks in their mouths for feeding (Figure 10.19). During the summer months, they survive as a thick-walled pupa.

This pest is particularly damaging in prolonged wet periods when the roots of young or succulent crops may be killed off. Occasionally lower leaves may be eaten.

Groundsmen sometimes control this pest by placing tarpaulins over water-soaked ground. The following morning the emerging leatherjackets may be brushed up. **Residual** chemicals, e.g. HCH, may be drenched or dusted into soil to reduce the larval numbers. Crops sown in autumn are rarely affected, as the larva is very small at this time.

***Sciarid fly*** (*Bradysia* sp.). The larvae of this pest (sometimes called **fungus gnat**) feed on fine roots of greenhouse pot plants, e.g. cyclamen, orchid, freesia, causing the plants to wilt. Fungus strands of mushrooms may be attacked in the compost. The slender black **females**, which are about 3 mm long, fly to a suitable site (freshly steamed compost, moss on sand benches, and well-fertilized compost containing growing plants), where 100 minute eggs are laid. The emerging legless larvae are white with a **black** head, and during the next month grow to a length of 3 mm before briefly pupating and starting the next life cycle.

The thin cuticle of the larva is penetrated by **residual** chemicals such as diazinon, applied as a drench, or incorporated into compost as a dust.

**Figure 10.19** *Leatherjacket larva (Courtesy Plant Pathology Laboratory, MAFF).*

The adults are excluded from mushroom houses by means of fine mesh screens placed next to ventilator fans, and by sprays of **fumigant** chemicals, e.g. dichlorvos.

### Beetles

This order of insects (Coleoptera) is characterized in the adult by hard, horny forewings which, when folded, cover the delicate hindwings used for flight. Most beetles are beneficial, helping in breakdown of humus, e.g. dung beetles, or feeding on pest species (*see* ground beetle). A few, e.g. wireworm, raspberry beetle, and vine weevil, may cause crop damage.

*Wireworm* (*Agriotes lineatus*). This beetle species is commonly found in grassland, but will attack most crops. The 1 cm long adult (click beetle) is brown-black and has the unusual habit of flicking itself in the air when placed on its back. The female lays eggs in weedy ground in May and June and these, after hatching, develop over a four-year period into slender 2.5 cm long, wireworm larvae (Figure 10.20), shiny golden-brown in colour, and possessing short legs. After a three-week pupation period in the soil, usually in

summer, the adult emerges, and in this stage survives the winter.

Turf grass may be eaten away by wireworms to show dry areas of grass. The pest also bores through potatoes to produce a characteristic tunnel, while in onions, brassicas and strawberries the roots are eaten away. In tomatoes, the larvae bore into the hollow stem.

Serious damage to young crops may be prevented by **seed dressings** with a **residual** chemical, e.g. HCH, which penetrates the thick insect cuticle. In greenhouses a drench of this chemical can be applied. Tomatoes, however, are sensitive.

*Raspberry bettle* (*Byturus tomentosus*). The developing fruit of raspberry, loganberry, and blackberry may be eaten away by the 8 mm long, golden-brown larvae of this pest (Figure 10.21). Only one life cycle per year occurs, the larva descending to the soil in July and August, pupating

**Figure 10.20** *Wireworm larvae on the soil surface (Courtesy Plant Pathology Laboratory, MAFF).*

**Figure 10.21** *Raspberry beetle larva eating a raspberry fruit (Courtesy Shell Chemicals).*

in a cell in which the golden-brown adult emerges and spends the winter before laying eggs in the host flower the next June.

Since the destructive **larval** stage may enter the host fruit and thus escape insecticidal control, the timing of the spray is vital. In raspberries, a **contact** chemical, e.g. malathion, applied when the fruit is pink, will achieve good control.

*Vine weevils* (*Otiorhyncus sulcatus*) belong to the beetle group, but possess a longer snout on their heads than other beetles. This species is 9 mm long, black in colour, with a rough textured cuticle. The forewings are fused together, making the pest incapable of flight. No males are known. The females lay eggs in August and September in the soil or compost. The emerging larvae are white, legless, and with a characteristic chestnut-brown head (Figures 10.5 and 10.22). They reach 1 cm in length in December when they **pupate** in the soil before developing into the adult.

The larva stage is the most damaging, eating away roots of crops such as cyclamen and begonias in greenhouses, primulas, strawberries, young conifers and vines outdoors. The adults may eat out neat holes in the foliage of hosts, e.g. rhodo-dendron and raspberry. Several related species,

**Figure 10.22** *Vine weevil adult (Courtesy Plant Pathology Laboratory, MAFF).*

e.g. the clay-coloured weevil (*Otiorhyncus singularis*) cause similar damage to that of the vine weevil.

Traps of corrugated paper placed near infested crops have achieved successful control in outdoor crops, while a **residual** chemical, e.g. HCH, may be drenched into the soil or incorporated in potting compost for control in greenhouse crops.

**Flea beetles** on leaves of cruciferous plants (e.g. stocks, cabbages) and **chafer larvae** on roots of turf in sandy areas are two other important beetle pests.

### Sawflies and bees

This group which, together with wasps and ants classified in the order Hymenoptera, are characterized by adults with two pairs of translucent wings, the fore and hind wings being locked together by fine hooks. The slender first segments of the abdomen give these insects a characteristic appearance.

*Rose leaf-rolling sawfly* (*Blennocampa pusilla*). The black shiny adults, resembling winged queen ants, emerge from the soil borne pupa in May and early June. Eggs are inserted into the leaf lamina which, in responding to the pest, rolls up tightly (Figure 10.23).

The emerging larva, which is pale green with a white or brown head, feeds on the rolled foliage and reaches a length of 1 cm by August, when it descends to the ground and forms an underground cocoon to survive the winter and pupate in March.

All types of roses are affected, although climbing roses are preferred. Damage caused by leaf-rolling tortrix caterpillars, e.g. *Cacoecia oparana*, may be confused with the sawfly, although the leaves are less curled.

Control is achieved, where necessary, by an application of nicotine dust which, on hot days in May and June, has a good **fumigant** action.

*Bees*. The well-known social insect, the honey-bee (*Apis mellifera*) is helpful to horticulturists. The female worker collects pollen and nectar in special pockets (honey baskets) on its hind legs.

**Figure 10.23** *Rose leaf-rolling sawfly damage (Courtesy Plant Pathology Laboratory, MAFF).*

This is a supply of food for the hive and, in collecting it, the bee transfers pollen from plant to plant. Several crops, e.g. apple and pear, do not set fruit when self-pollinated. Thus, this insect provides a useful function to the fruit grower. In large areas of fruit production the number of resident hives is insufficient to provide effective pollination, and in cool, damp, or windy springs, the flying periods of the bees are reduced. It may therefore be advantageous for the grower to introduce bee hives into the orchards during the blossom time, as an insurance against bad weather. One hive is normally adequate to serve 0.5 hectare of fruit. Blocks of four hives placed in the centre of a 2 hectare area require foraging bees to travel a maximum distance of 140 metres.

In addition to honey bees, wild species, e.g. the potter flower bee (*Anthophora retusa*) and red-tailed bumble-bee (*Bombus lapidarius*) increase fruit set, but their numbers are not really high enough to dispense with the honey bee. The pollination of tomatoes in greenhouses is commonly achieved by small in-house populations of a bumble-bee species (*Bombus terrestris*). All species of bee are killed by broad spectrum insec-

ticides, e.g. azinphos-methyl, and it is important that spraying of such chemicals be restricted to early morning or evening during the blossom time period when hives have been introduced.

## MITES

**This important group (Acarina), classified with spiders and scorpions in the Arachnida, is distinguished from insects by the possession of four pairs of legs, a fused body structure, and by the absence of wings (Figure 10.24). Many of the tiny soil-inhabiting mites serve a useful purpose in breaking down plant debris.**

Several above-ground species are serious pests on plants. The **life cycle** is composed of egg, larva, nymph, and adult.

*Glasshouse red spider mite* (*Tetranychus urticae* and *T. cinnabarinus*). These pests are of tropical origin, and thrive best in high greenhouse temperatures. The first species is 1 mm long, yellowish in colour, with two black spots (Figure 10.23). The female lays about 100 tiny spherical eggs on the underside of the leaf, and after a period of three days the tiny six-legged larva moults to produce the nymph stage which resembles the adult. The life cycle length varies markedly from sixty-two days at 10°C, to six days at 35°C. The pest's multiplication potential is extremely high. In autumn, when the daylight period decreases to fourteen hours and temperatures fall, egg production ceases and the fertilized females, which are now red in colour, move into the greenhouse structures to hibernate (diapause), representing foci for next spring's infestation. The **second** species, which is dark reddish-brown, has a similar life cycle to the first, but does not hibernate. The first species is common on annual crops such as tomatoes, cucumbers, chrysanthemums, while the second species is more common on the perennial crops such as carnations, arums and hot-house pot plants. The two species often occur together on summer hosts.

Because the piercing mouthparts inject poisonous secretions, the mites cause localized death of leaf mesophyll cells. This results in a fine mottling on the leaf (Figure 10.25), not to be confused with the larger spots caused by **thrips**. In large numbers the mites can kill off leaves and eventually whole plants. Fine silk strands are produced in severe infestations, appearing as 'ropes' on which the mites move down the plant. On chrysanthemums, these ropes make the plant unsaleable.

Control may be achieved in three ways. A predatory mite, *Phytoseiulus persimilis*, is commonly introduced into cucumber, chrysanthemum and tomato crops in spring. Winter **fumigation** of greenhouse structures with chemicals such as formalin or burning sulphur kill off some of the hibernating females. A range of **chemicals**, e.g. dicofol (**contact** action), derris (**stomach** action), and aldicarb (**systemic** action) are applied to crops against this pest, but care must be taken not to cause chemical scorch in plants, not to use chemicals to which the pest has become resistant, and not to kill predators by unthinking use of these chemicals (*see* integrated control).

**Gall mite of blackcurrant** (*Cecidophyopsis ribis*). Unlike red spider mite, this species, sometimes called **big bud mite**, is elongated in shape, and minute (0.25 mm) in size. It spends most of the year living inside the buds of blackcurrants and, to a lesser extent, other *Ribes* species. Breeding takes place inside the buds from June to September, and January to April. In May the mites emerge and disperse on silk threads and on the bodies of aphids. The damaged bud meristem produces many scale leaves which gives the bud its unusual appearance (Figure 10.26). These buds often fail to open, or produce distorted leaves.

The mite carries the virus-like agent responsible for the damaging **reversion disease**, which stunts the plant and reduces fruit production.

The mite is controlled in three ways. **Clean planting material** is essential for the establishment of a healthy crop. **Pruning** out of stems with big bud and destruction of reversion-infected plants slow down the progress of the pest. Since no chemicals can control the mite in the bud, sprays of contact **acaricides**, e.g. endosulfan, are applied in the May–June period when the mites are migrating.

**Tarsonemid mite** (*Tarsonemus pallidus*). This spherical mite, only 0.25 mm in length, lives in the unexpanded buds of a wide variety of pot plants, e.g. *Amaranthus*, fuchsia, pelargonium and cyclamen, and is often called the cyclamen mite. A closely related but distinct strain is found on

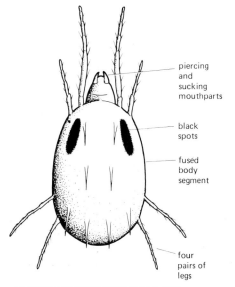

piercing and sucking mouthparts

black spots

fused body segment

four pairs of legs

GLASSHOUSE RED SPIDER MITE (x 100)

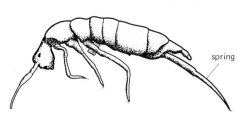

spring

SPRINGTAIL (x 40)

**Figure 10.24** *Glasshouse red spider mite and springtail insect. Note that the red spider mite may be light green or red in colour. Its extremely small size (0.8 mm) enables it to escape a grower's attention. The springtail (about 2 mm in size) can jump by means of a spring at the end of its body.*

**Figure 10.25** *Glasshouse red spider mite damage: see also Figures 10.24 and 12.1.*

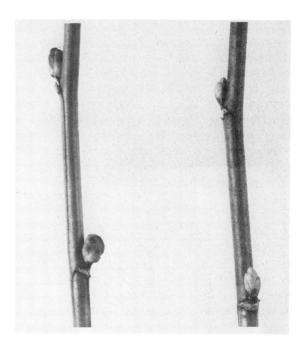

**Figure 10.26** *Blackcurrant gall mite damage: note the 'big-bud' symptoms on the left compared with the normal buds on the right (Courtesy Plant Pathology Laboratory, MAFF).*

strawberries. In greenhouses, the adults may lay eggs all the year round, and the **two-week** life cycle period can cause a rapid increase in its numbers.

The small feeding holes and injected poisons from the mite mouthparts combine to distort the developing leaf and flower buds of the affected crop to such an extent that leaves and petals are stunted and misshapen, and flowers may not open properly.

Care should be taken to prevent introduction of infested plants and propagative material into greenhouses. Contact acaricides, e.g. endosulfan and dicofol, are effective against the mite. Addition of a **wetter/spreader** may help the spray penetrate the tight-knit scale leaves of the buds.

Four other horticulturally important mites require a mention. The fruit tree red spider mite (*Panonychus ulmi*) causes serious leaf mottling of the ornamental *Malus* and apple. Conifer spinning mite (*Oligonychus ununguis*) causes spruce to yellow, and spins a web of silk threads. Bulb scale mite (*Steneotarsonemus laticeps*) causes internal discoloration of forced narcissus bulbs. Bryobia mite (e.g. *Bryobia rubrioculus*) attacks fruit trees,

and may cause damage to greenhouse crops, e.g. cucumbers, if blown in from neighbouring trees.

## OTHER ARTHROPODS

In addition to insects and mites, the phylum Arthropoda contains three horticulturally relevant classes, the Crustacea (woodlice), Symphyla (symphilids), and Diplopoda (millipedes).

*Woodlouse* (*Armadillidium nasutum*), a relative of the marine crabs and lobsters, has adapted for terrestrial life, but still requires damp conditions to survive. In damp soils it may number over a million per hectare, and greatly helps the breakdown of plant debris, as do earthworms. In greenhouses, where plants are grown in hot, humid conditions, this species may multiply rapidly, producing two batches of 50 eggs per

**Figure 10.27** *Woodhouse.*

year. The adults (Figure 10.27) roll into a ball when disturbed.

The damage is confined mainly to stems and lower leaves of cucumbers, but occasionally young transplants may be nipped off near ground level, in the manner of cutworms and leather-jacket damage.

Partial soil **sterilization** by steam or chemicals, e.g. methyl bromide, effectively controls wood-lice. Rotting brickwork and timber provide refuge for them, and should be replaced. **Residual** chemicals, e.g. HCH, may be incorporated or drenched into the growing medium.

*Symphilids* (*Scutigerella immaculata*). These delicate white creatures, with **twelve** pairs of legs, resemble small millipedes. The adult female, 6 mm long, lays eggs in the soil all the year round, and the development through larvae to the adult takes about three months. Symphilids may migrate two metres down into soil during hot, dry weather.

In greenhouse crops the **root hairs** are removed, and may cause lettuce to mature without a heart. Infectious fungi, e.g. *Botrytis*, may enter the roots after symphilid damage. The recognition of this pest is made easier by dipping a suspect root and surrounding soil into a bucket of water and searching for organisms floating on the water surface.

Injection of **fumigant** chemicals, e.g. metham-sodium, into greenhouse soils is an effective method of control against this deep-living pest. Steam sterilization fails because the pest can crawl ahead of the steam's advance. Drenches of a

**residual** chemical, e.g. diazinon or HCH, may be necessary with severe attacks in both greenhouse and outdoor crops.

*Millipedes*. These elongated, slow-moving creatures are characterized by a thick cuticle, and the possession of many legs, two pairs to each body segment (Figure 10.28). Many species are useful in breaking down soil organic matter (*see* arthropods), but two pest species, the flat mil-lipede (*Brachydesmus superus*) and a tropical species (*Oxidus gracilus*), cause damage to roots of strawberries and cucumbers respectively.

Control of these pests is difficult, although residual chemicals, e.g. HCH, have some effect. Fortunately the slow rate of development of these creatures makes control largely unnecessary.

*Centipedes* resemble millipedes, but are much more active. Searching for insects and worms in the soil (Figure 10.29), they may help control soil pests.

## NEMATODES

This group of organisms, also called **eelworms**, is found in almost every part of the terrestrial environment, ranging in size from the large animal

**Figure 10.28** *Millipede: this group of animals has two pairs of legs to each segment.*

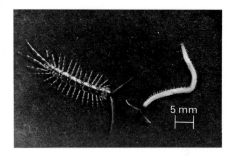

**Figure 10.29** *Centipedes: two dissimilar species are shown.*

parasites, e.g. *Ascaris* (about 20 cm long) in livestock, to the tiny soil-inhabiting species (about 0.5 mm long). Non-parasitic species may be beneficial, feeding on plant remains and soil bacteria, and helping in the formation of **humus**. The general structure of the nematode body is shown in Figure 10.30. A feature of the plant parasitic species is the **spear** in the mouth region, which is thrust into plant cells. Salivary enzymes are then injected into the plant and the plant juices sucked into the nematode (Figure 10.31). Nematodes are very active animals, moving in a wriggling fashion in soil moisture films, most actively when the soil is at **field capacity**, and more slowly as the soil

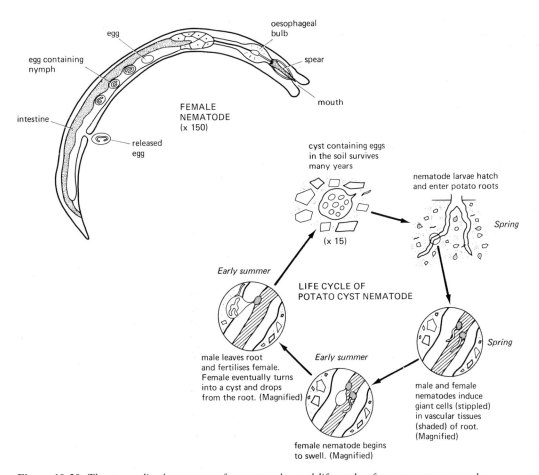

**Figure 10.30** *The generalized structure of a nematode, and life cycle of potato cyst nematode.*

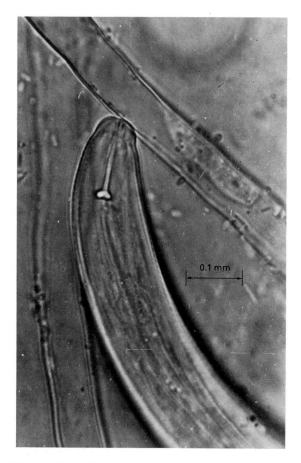

0.1 mm

**Figure 10.31** *Nematode feeding: note the spear inside the mouth, used to penetrate plant tissues (Courtesy J. Bridge).*

either waterlogs or dries out. Five horticulturally important types are described below.

*Potato cyst nematode* (*Globodera rostochiensis* and *G. pallida*). This serious pest is found in most soils that have grown potatoes. A proportion of the eggs in the soil hatch in spring, stimulated by chemicals produced in potato roots. The larvae invade the roots, disturbing **translocation** in xylem and phloem tissues, and sucking up plant cell contents. When the adult male and female nematodes have developed, they migrate to the outside of the root, and the now swollen female leaves only her head inserted in the plant tissues. After fertilization, the white female becomes

almost spherical, about 0.5 mm in size, and contains 200–600 eggs (Figure 10.30). As the potato crop reaches harvest, the female changes colour. In *G. rostochiensis* (the golden nematode), the change is from white to **yellow**, and then to dark brown, while in the other species, *G. pallida*, no yellow phase is seen. The significance of the species' differences is seen later. Eventually the dark brown female dies and falls into the soil. This stage, which looks like a minute onion, is called the **cyst**, and the eggs inside this protective shell may survive for ten years or more.

The pest may be diagnosed in the field by the mature white or yellow females seen on the potato roots. Leaves show a yellowing symptom, plants are often severely stunted, and occasionally killed. The distribution of damage in a field is characteristically in patches. Tomatoes grown in greenhouses and outdoors may be similarly affected.

Several forms of control are available against this pest. Since it attacks only potatoes and tomatoes, **rotation** is a reliable if inconvenient way of overcoming the problem. Since it is known that an average soil population of ten cysts per 100 gm of soil results in a three tonne per hectare decline in yield, a soil count for cysts can indicate to a grower whether a field should be used for a potato crop. Early potatoes are lifted before most nematodes have reached the cyst stage, and thus escape serious damage. Some potato cultivars, e.g. 'Pentland Javelin' and 'Maris Piper', are resistant to golden nematode strains found in Great Britain, but not to *G. pallida*. Since the golden nematode is dominant in the south of England, use of resistant cultivars has proved effective in this region. **Residual** chemicals (e.g. aldicarb), incorporated into the soil at planting time, provide economical control when the nematode levels are moderate to fairly high, but are not recommended at low levels because they are not economic, or at high levels because the chemical kills insufficient nematodes.

*Stem and bulb eelworm* (*Ditylenchus dipsaci*) attacks many plants, e.g narcissus, onions, beans and strawberries. Several strains are known, but their host ranges are not fully defined. The 1 mm long nematodes enter plant material and breed continuously, often with thousands of individuals in one plant. When an infected plant matures, the nematodes dry out in large numbers, appearing as white fluffy **eelworm wool** which may survive for several years in the soil. **Weeds**, such as bindweed, chickweed, and speedwells, act as alternate hosts to the pest.

The damage caused by this species varies with the crop attacked. Onions show a loose puffy appearance (called bloat); carrots have a dry mealy rot; the stems of beans are swollen and distorted. Narcissus bulb scales show brown rings when cut across (Figure 10.32). Their leaves show raised yellow streaks and the crop flowers late.

Control is achieved in several ways. Control of weeds (*see* chickweed); **rotation** with resistant crops, e.g. lettuce and brassicas; use of clean, nematode-free **seed** in onions; **hot water** treatment of onions and narcissus at precisely controlled temperatures; and incorporation of **residual** granules, e.g. oxamyl, incorporated into soils at the time of planting onions; all these methods help reduce this serious pest.

***Chrysanthemum eelworm*** (*Aphelenchoides ritzemabosi*). It has been seen that some nematodes live in soil (e.g. cyst eelworms), others move into stems (e.g. stem eelworms). This 1 mm long nematode spends most of its life cycle in young leaves of crops, e.g. chrysanthemum, *Saintpaulia* and strawberries. The adults move along films of water on the surface of the plant, and enter the leaf through the stomata. They breed rapidly, the females laying about thirty eggs, which complete a life cycle in fourteen days. During the winter they live as adults in stem tissues, but very few overwinter in the soil.

The first symptom is blotching and purpling of the leaves, which spreads to become a dead brown area between the veins. The lower leaves are worst affected. When buds are infested, the resulting leaves may be misshapen. Greenhouse-grown chrysanthemums are rarely affected, as they are raised from pest-free cuttings.

There are two main forms of control. **Warm water treatment** of dormant chrysanthemum stools, e.g. at 46°C for five minutes, is very effective for outdoor grown plants. Control by application of a **granular systemic** chemical, e.g. aldicarb to pot plants or beds, is also effective.

***Root knot eelworm*** (*Meloidogyne* spp.) found mainly in greenhouses, is of tropical origin, and thrives in high temperature conditions, causing typical galls, up to 4 cm in size (Figure 10.33), on the roots of plants such as chrysanthemum, *Begonia*, cucumber and tomato. The swollen female lays 300–1000 eggs inside the root and on the root surface. These eggs can survive in root debris for over a year, and are an important source of subsequent infestations. The larvae

**Figure 10.32** *Stem and bulb nematode symptoms in a daffodil bulb: note the dark tissue between the bulb scales (Courtesy Plant Pathology Laboratory, MAFF).*

**Figure 10.33** *Root knot nematode damage on cucumber: note the enlarged roots (Courtesy C.C. Doncaster).*

hatch from the eggs and search for roots, reaching soil depths of 40 cm, surviving in damp soil for several months. On entering the plant, the nematode larva stimulates the adjoining root cells to enlarge. These cells block movement of water to the root stele (*see* root structure), which results in the wilting symptoms so commonly seen with this pest.

Care should be taken not to transfer infested soil with transplants from one greenhouse to another. **Steam sterilization** effectively controls the nematode only if the soil temperature reaches 99°C to a depth of 45 cm. Less stringent sterilization often results in a severe infestation in the next crop. **Chemical sterilization** of soil with fumigant chemicals, e.g. dichlorpropene and methyl bromide, is effective if a damp seedbed tilth is first prepared. **Resistant** tomato rootstocks, e.g. KVNF, have been used on grafted plants. A 2.5 cm layer of peat placed around roots of infested plants allows new root growth. **Nutrient film** and soil-less methods of growing reduce the pest's likely importance in a crop.

*Migratory plant nematodes*. The four species of nematodes previously described spend most of their life cycle inside plant tissues (endoparasites). Some species, however, feed from the outside of the root (ectoparasites). The **dagger** nematodes (e.g. *Xiphinema diversicaudatum*) and **needle** nematodes (e.g. *Longidorus elongatus*), which reach lengths of 0.4 cm and 1.0 cm respectively, attack the young roots of crops such as rose, raspberry and strawberry, and cause stunted growth. In addition, these species transmit the important **viruses**, arabis mosaic on strawberry, and tomato black ring on ornamental cherries. The nematodes may survive on the roots of a wide variety of weeds. Control is achieved in fallow soils by the injection of a fumigant chemical, e.g. dichloropropene or dazomet.

## PRACTICAL EXERCISES

1   Observe a large slug eating plant material. Watch its movement across a glass plate by viewing from underneath the glass.
2   Place some leaf litter in a tray and, using a hand lens, observe the harmful pests, springtails and symphilids, and the useful insects, mites and nematodes.
3   Compare the mouthparts of the **sucking** aphid and **biting** caterpillar, as seen through a hand lens.
4   Look for a severely stunted potato crop. Observe the root system for potato cyst nematode cysts.

## FURTHER READING

Alford, D.V. *A Colour Atlas of Fruit Pests* (Wolfe Science, 1984).
Alford, D.V. *Pests of Ornamental trees, Shrubs and Flowers* (Wolfe Publishing, 1991).
British Crop Protection Council. *The UK Pesticide*

*Guide* (1993).

Buczacki, S., and Harris, K. *Guide to Pests, Diseases and Disorders of Garden Plants* (Collins, 1992).

Dahl, M.H., and Thygesen, T.B. *Garden Pests and Diseases* (Blandford Press, 1974).

Gratwick, M. *Crop Pests in the UK* (Chapman & Hall, 1992).

Hope, F. *Turf Culture: A Manual for Groundsmen* (Cassell, 1990).

Jones, F.G.W., and Jones, M.G. *Pests of Field Crops* (Edward Arnold, 1984).

Morgan, W.M., and Ledieu, M.S. *Pests and Disease Control in Glasshouse Crops* (British Crop Protection Council, 1979).

Pirone, P.P. *Diseases and Pests of Ornamental Plants* (Wiley, 1985).

Savigear, E. *Garden Pests and Predators* (Blandford, 1992).

Scopes, N., and Stables, L. *Pests and Disease Control Handbook* (British Crop Protection Council, 1989).

# 11

# Fungi, Bacteria and Viruses

*In this chapter the three groups of organisms are described, and emphasis is placed on the damage they cause, the important aspects of their life cycles, and the control measures which are available to reduce their infection and spread. Within each group, the diseases are classified according to the part of the plant attacked. Because of the microscopic size of the fungal spore, bacterial cell or virus particle, it may be difficult for the grower to relate the causitive* **organism** *to the disease as he sees it. For example,* Phytophthora infestans, *a fungal organism, travels through the air as a spore to infect* **potato** *leaves. When the leaves start to turn black, the plant is said to show signs or* **symptoms** *of the* **potato blight** *disease. Only when the plant's growth and tuber production are decreased by the organism is* **yield loss** *said to occur. This chapter emphasizes the fact that symptoms of a disease are the horticulturist's main guide to the presence of fungi, bacteria and viruses.*

## FUNGI

### Structure and biology

**These organisms, commonly called moulds, cause serious losses in all areas of horticulture. They are** thought to have common ancestors with the filamentous algae, a group including the present-day green slime in ponds.

**The fungus is composed in most species of microscopic strands (hyphae), which may occur together in a loose structure (mycelium), form dense resting bodies (sclerotia), (Figure 11.1) or produce complex undergound strands** (*see* rhizomorphs).

The hyphae in most fungi are capable of producing spores. Wind-borne spores are generally very small (about 0.01 mm), not sticky, and often borne by hyphae protruding above the leaf surface, e.g. **grey mould**, so that they catch turbulent wind currents. Water or rain-borne spores are often sticky, e.g. **damping-off**. **Asexual** spores produced without fusion of two hyphae commonly occur in seasons favourable for disease increase, e.g. humid weather for downy mildews; dry, hot weather for powdery mildews. **Sexual** spores, produced after hyphal fusion, commonly develop in unfavourable conditions, e.g. a cold, damp autumn, and they may be produced singly as in the downy mildews, or in groups within a protective hyphal **spore case**, often observable to the naked eye, as in the powdery mildews. Different genera and species are identified by microscopic measurement of the shape and size of the sexual spores, but when sexual spores are absent, asexual spores or hyphae are used for identification. Horticulturists without microscopes must use the symptoms as a guide to the cause of the disease. While disease causing or **parasitic** fungi are the main

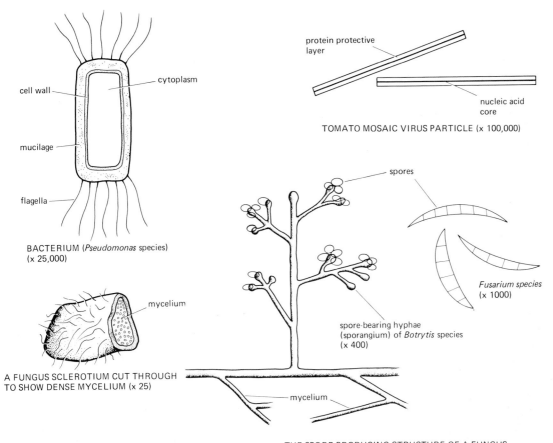

**Figure 11.1** *Microscopic details of a virus, bacterium and three fungi. Note the relative sizes of the organisms.*

concern of this chapter, in many parts of the environment there are **saprophytic** fungi which break down organic matter (*see* Chapter 15), and **symbiotic** fungi that may live in close association with the plant, e.g. mycorrhiza fungi in fine roots of conifers.

The spore of a leaf-infecting fungal parasite, after landing on the leaf, produces a germination tube which, being delicate and easily dried out, must enter through the **cuticle** or **stomata** within a few hours before dry, unfavourable conditions recur. Within the leaf, the hyphae grow, absorbing food until, within a period of a few weeks, they produce a further crop of spores (Figure 11.2). Leaf diseases such as potato blight often increase

very rapidly when conditions are favourable. Roots may be infected by spores, e.g. in damping off; hyphae, e.g. wilt diseases; sclerotia, e.g. *Sclerotinia* rot; or rhizomorphs, e.g. honey fungus. Root diseases are generally less affected by short periods of unfavourable conditions and often increase at a constant rate.

**Leaf and flower diseases**

**Downy mildew of cabbage and related plants** (*Perenospora brassicae*)

This serious disease causes a white bloom mainly

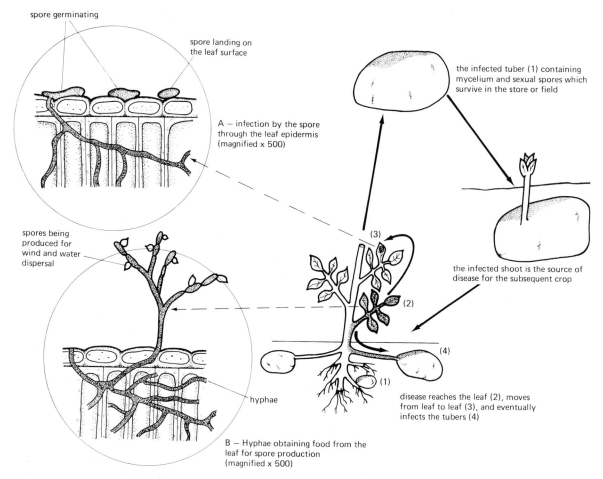

spore germinating

spore landing on
the leaf surface

the infected tuber (1) containing
mycelium and sexual spores which
survive in the store or field

A — infection by the spore
through the leaf epidermis
(magnified x 500)

spores being
produced for
wind and water
dispersal

the infected shoot is the source of
disease for the subsequent crop

hyphae

disease reaches the leaf (2), moves
from leaf to leaf (3), and eventually
infects the tubers (4)

B — Hyphae obtaining food from the
leaf for spore production
(magnified x 500)

**Figure 11.2** *Infection and life cycle of potato blight fungus. The left side illustrates microscopic infection of the leaf. The right side shows how the disease survives and spreads.*

on the **under surface** of leaves of ornamental cruciferous plants such as stocks and wallflowers, on brassicas, and occasionally on weeds such as shepherd's purse. The disease is most damaging when seedlings are germinating, particularly in spring when the young infectable tissues of the host plant and favourable damp conditions may combine to kill off a large proportion of the developing plants. Other crops such as lettuce and onions are attacked by different downy mildews, *Bremia lactucae* and *Perenospora destructor* respectively, and no cross infection is seen between different unrelated crops.

Asexual spores (zoospores) are produced mainly in spring and summer, and a spray of a protective chemical (e.g. **dichlofluanid**) is commonly used at the seedling stage to kill off spores on the leaf.

Thick-walled sexual spores (oospores) produced within the leaf tissues fall to the ground with the death of the leaf, survive the winter and initiate the spring infections. It is thus not advisable to grow successive brassicas in the same field, and particularly not to sow in spring next to overwintered crops.

**Potato blight** (*Phytophthora infestans*)

This important disease is a constant threat to potato production, and caused the Irish potato famine in the nineteenth century. The first symptoms seen in the field are yellowing of the foliage, which quickly goes **black** and then produces a white bloom on the under surface of the leaf in damp weather. The stems may then go black, killing off the whole plant. The tubers may show dark surface spots which, internally, appear as a **deep dry red-brown** rot. This fungus may attack tomatoes, the most notable symptom being the **dark brown blisters** on the fruit.

The fungus survives the winter as mycelium and sexual spores in the tubers (Figure 11.2). The spring emergence of infected shoots results in the production of asexual spores which, when carried by wind, land on potato leaves or stems and can, after infection, result in a further crop of spores within a few days under warm, wet weather conditions. Thus the disease can spread very quickly. Later in the crop, badly infected plants cause tuber infection as rainfall may wash down spores into the soil.

Several **preventative** control measures are used (*see* removal of material). Clean 'seed' lowers infection within neighbouring crops. Removal and herbicidal destruction (e.g. dichlobenil) of diseased tubers from stores or 'clamps' similarly prevent disease spread. Knowledge of the disease's moisture requirement leads to better control. By measuring both the duration of atmospheric relative humidity greater than 92%, and the mean daily temperatures greater than 10°C (together known as **critical periods**), forecasts of potato blight outbreaks may be made and protectant sprays of chemicals such as **zineb** applied before infection takes place. Resistant potato cultivars prevent rapid build-up of disease, although resistance may be overcome by newly occurring fungal strains. Early potato cultivars usually complete tuber production before serious blight attacks, while maincrop top growth may deliberately be killed off with foliage-acting herbicides such as **diquat** to prevent disease spread to tubers.

A curative control measure employs a systemic fungicide, e.g. **metalaxyl**, which penetrates the leaf and kills the infecting mycelium. Such chemicals must be mixed with a protectant ingredient to reduce the development of fungicidal resistance.

**Powdery mildew of ornamental *Malus* and apple** (*Podosphaera leucotricha*)

Powdery mildews should not be confused with downy mildews. This disease is distinguished by its dry powdery appearance, most commonly found on the **upper** surface of the leaf, and by its preference for **hot**, **dry** weather conditions.

The disease survives the winter as mycelium within the buds, which often appear small, and the infected twigs have a dried, **silvery** appearance. The emergence of the mycelium with the germinating buds in spring results in a white bloom over the young leaves (**primary** mildew), which may be reduced by a **winter** application of a selective detergent-like fungicide sprayed before bud burst, or by winter pruning. As the spring progresses, chains of asexual spores produced on the outside of the leaf are carried by wind and cause the destructive *secondary* mildew which, growing **externally** over the leaf surface, sucks out the leaf's moisture and may cause premature leaf drop. Flowering normally occurs before the secondary infection stage, but infection in young fruit may produce a rough skin (**russeting**). Other species of fruit such as pears, quinces, medlars and ornamental *Malus* may be affected by this organism. Other species of powdery mildew commonly occur in horticulture: *Sphaerotheca pannosa* on rose, and *S. fuliginea* on cucumber. Cross infection between these crops does not occur.

Sexual spores may be produced inside a dark coloured **spore case** (cleistothecia), about 1 mm in size in autumn. Though not important in horticulture as an overwintering stage, it may assume a vital role in powdery mildew of cereals.

Control is achieved by preventative measures mentioned above, and by curative spring and summer sprays, using a wide choice of active ingredients which may act in a contact manner on the external mycelium e.g. **dinocap** (*see* protectant), or in a systemic manner on the internal feeding hyphae (haustoria), e.g. **benomyl** (*see* systemic).

**Black spot of roses** (*Diplocarpon rosae*)

This common disease in gardens and greenhouse-produced roses is first seen as dark leaf spots, which may be followed by general leaf yellowing and then leaf drop (Figure 11.3). The infection of young shoots has a slow weakening effect on the whole plant.

**Asexual** spores produced within the leaves are released in wet and mainly warm weather conditions, and are then carried a limited distance by rain drops or irrigation water before beginning the cycle of infection again. No overwintering sexual stage is seen in Britain, and it is probable that asexual spores surviving in autumn produced wood or in fallen leaves begin the infection process the following spring. Control is difficult, as resistance in rose cultivars is not common, and vigorous growth in late spring and summer prevent continuous protective control by chemicals, e.g. **captan**. The addition of a **wetter/spreader** may improve control by spreading the active ingredient more effectively over the leaf surface. In industrial areas the sulphur dioxide in the air may be at sufficient concentration to reduce black spot.

**Carnation rust** (*Uromyces dianthi*)

The rust fungi are a distinct group which may

**Figure 11.3** *Rose black spot: a healthy leaf is shown on the left.*

produce five spore forms within the same species. When the spore forms occur on more than one host, e.g. blackcurrant rust (*Cronartium ribicola*), which attacks both blackcurrants and five-needle pines, the close association of the two crops may give rise to high rust levels. Most horticultural rust species are found on only **one** host species.

Carnation rust first appears as an indistinct **yellowing** of the leaf and stem, soon turning to an **elongated raised brown** spot which yields brown dust (spores) when rubbed (Figure 11.4).

The more common thin-walled spores (uredospores) are spread by wind currents, and infect the leaf by way of the stomata in damp conditions. The less common thick-walled spores (teleutospores) may survive and overwinter in the soil. The resistance of carnation cultivars varies, while related species, e.g. pinks or sweet williams, are rarely affected.

**Preventative** control includes the use of rust-free cuttings, sterilization of border soils, and careful maintenance of greenhouse ventilators to prevent damp patches occurring in the crop. **Chemical** control is achieved by **protectant** (e.g. zineb), or **systemic** (e.g. benodanil) sprays.

The occurrence of white rust (*Puccinia horiana*) on chrysanthemums has created serious problems for the industry and for gardeners, because of the ease with which the disease is carried in cuttings, and because of its speed of increase and spread.

**Figure 11.4** *Carnation rust: note the dark rusty lesions on both leaf and stem (Courtesy Glasshouse Crops Research Institute).*

### Stem diseases

**Grey mould** (*Botrytis cinerea*)

This disease is most commonly recognized by the dense, light grey fungal mass which follows its infection. In lettuce, the whole plant rots off at the base and the plant goes yellow and dies. In tomatoes, infection in damaged side shoots, and yellow spots (ghost spots) on the unripe and ripe fruit are found. In many flower crops, e.g. chrysanthemums, infected petals show purple spots which, in very damp conditions, lead to a mummified flower head. Most crops may be affected by this disease.

Grey mould requires **wounded tissue** for infection, which explains its importance in crops which are deleafed, e.g. tomatoes, or disbudded, e.g. chrysanthemums. Damp conditions are essential for its infection and spore production. The millions of spores are carried by wind to the next wounded surface. Black **sclerotia**, about 2 mm across, produced in badly infected plants, often act as the overwintering stage of the disease after falling to the ground, and are particularly infective in unsterilized soils on young seedlings and delicate plants, e.g. lettuce.

**Preventative** control may involve soil steriliz-ation or soil incorporated dusts, e.g. **dicloran** used against the sclerotia. Strict attention to greenhouse humidity control, particularly over-night, limits the dew formation so important in the organism's infection. Removal of infected tis-sue is possible in sturdy plants, e.g. tomatoes, and a protective fungicide paste, e.g. **iprodione**, is often then applied to the cut surface. Protective sprays, e.g. **iprodione**, are often applied to green-house grown crops to prevent the spores germi-nating, and to reduce spore production.

**Apple canker** (*Nectria galligena*)

This fungus causes sunken areas in bark of both young and old branches of ornamental *Malus*, apples, or pears (Figure 11.5), and occasionally on fruit. Poor shoot growth is seen, and the wood may fracture in high winds. The cankers may bear red spore cases (**perithecia**), resembling red spider mite eggs in autumn.

The fungus enters through leaf scars in autumn or through pruning wounds during winter.

Care is therefore necessary to prevent infection,

**Figure 11.5** *Apple canker: note how the wood has been disfigured by the fungus.*

particularly in susceptible apple cultivars, e.g. 'Cox's Orange Pippin', by applying a proprietary paint containing, for example, **mercuric oxide**, to pruned surfaces soon after pruning. **Removal** of cankered shoots may be necessary to prevent further infection, while in cankers of large branches, cutting out of brown infected tissue may allow the branch to be used. Removed tissue should be burnt. Some growers apply a spray of **copper** (Bordeaux) at bud burst (spring) and leaf fall (autumn) to prevent entry of germinating spores.

**Dutch elm disease** (*Ceratocystis ulmi*)

The first symptom of this disease is a **yellowing** of foliage in one part of the tree in early summer. The foliage then dies off progressively from this area of the tree, often resulting in death within three months. Trees which survive one year's infection may fully recover in the following year. All common species and hybrids of elm growing in Great Britain are susceptible to the disease.

The causative fungus lives in the **xylem** tissues of the stem, and produces a poison which results in a blockage of the water-conducting vessels, causing the wilt which is observed. Two black and red wood-boring species of beetle, *Scolytus scolytus* (5 mm long), and *Scolytus multistriatus* (3 mm long), enter the stems, leaving character-istic 'shot holes'. Eggs are laid, and a **fan-shaped** pattern of galleries is produced under the bark by the larvae which later, as adults, emerge from the wood, carrying sticky asexual and sexual spores to continue the spread of the disease to other unin-fected elms. Graft transmission from tree to tree by roots commonly occurs in hedge grown elms.

The cost of preventative control on a large number of uninfected trees in uneconomic, though high pressure injection of a **systemic** fungicide, (e.g. benomyl salts), which travels upwards through the xylem tissues, has proved successful in some cases. Selections of **disease-tolerant** hybrids, e.g. *Ulmus x vegeta* 'Groenveld', may replace the common species and hybrids.

### Root diseases

**Club root** (*Plasmodiophora brassicae*)

This disease causes serious damage to most members of the Cruciferae family, which includes cabbage, cauliflowers, Brussels sprouts, stocks and *Alyssum*. Infected plants show signs of **wilting** and yellowing of older leaves, and often severe stunting. On examination, the roots appear **stubby** and **swollen** (Figure 11.6), and may show a wet rot.

The club root organism survives in the **soil** for more than five years as minute spores which germinate to infect the **root hairs** of susceptible

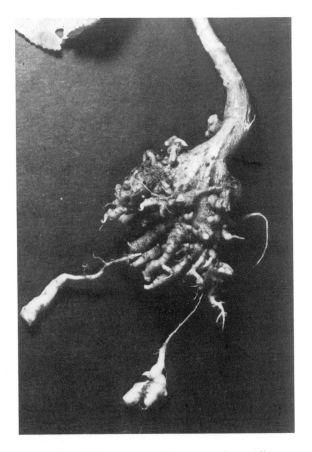

**Figure 11.6** *Club root on cabbage: note the swollen tap and side roots.*

plants. The fungus is unusual in forming a jelly-like mass (**plasmodium**), not hyphae, within the tissue. The plasmodium stimulates root cell division and causes cell enlargement, which produces swollen roots. The flow of food and nutrients in **phloem** and **xylem** is disturbed, with consequent poor growth of the plant. With plant maturity, the spores produced by the plasmodium within the root are released as the root rots. The disease is favoured by high soil moisture, high soil temperatures, and by acid soils.

Several **preventative** control measures may be used. **Rotation** helps by keeping cruciferous crops away from high spore levels. Autumn-sown plants establish in soil temperatures unfavourable to the disease, and are normally less infected. In transplanted crops, the use of a seed bed previously treated with a sterilant, e.g. **dazomet**, ensures healthy transplants. Liming of soil inhibits spore activity (*see* soil fertility). Compost made from infected brassica plants should be avoided. Transplanted brassicas may be dipped in a paste fungicide e.g. **thiophanate-methyl**.

**Damping off** (*Pythium and Phytophthora species*)

These two fungi cause considerable losses to the delicate **seedling** stage. The infection may occur below the soil surface, but most commonly the emerging seedling plumule is infected at the soil surface, causing it to topple. Occasionally the roots of mature plants, e.g. cucumbers, are infected, turn brown and soggy, and the plants die. Both *Pythium* and *Phytophthora* occur naturally in soils as **saprophytes**, and under damp conditions produce asexual spores which cause infection. Sexual spores (oospores) are produced in infected roots and may survive several months of dry or cold soil conditions.

Prevention control is best achieved against these diseases by sterilization of soil by heat, by chemicals, e.g. methyl bromide which kills off all stages of these fungi, or by coating the crop seed with a protective chemical (*see* seed dressing), e.g. captan, to prevent early infection. Water tanks with open tops, harbouring rotting leaves, are 'a common source of infected water and should be cleaned out regularly. Sand and capillary matting on benches in glasshouses should be regularly washed in hot water, and the use of door mats soaked in a fungicide, e.g. formalin, may prevent foot-spread of the organisms from one greenhouse to another. Water-logged soils should be avoided, as the fungi increase most rapidly under these conditions. Chemical prevention is achieved in some crops by the incorporation of a fungicide, e.g. **etridiazole** into compost, or by drenching with a fungicide not toxic to roots, e.g. **copper sulphate** and **ammonium carbonate** (Cheshunt) mixture.

**Conifer root rot** (*Phytophthora cinnamomi*)

This soil-inhibiting fungus causes the foliage of plants to turn **grey-green**, then brown, and eventually to die off completely. Sliced roots show a chestnut **brown rot**, with a clear line between infected and non-infected tissues. Two hundred plant species, including *Chaemaecyparis*, *Erica* and *Rhododendron* species may be badly attacked.

The disease is commonly introduced on infected stock plants, or contaminated footwear. It multiplies most rapidly under wet conditions, and within a temperature range of 20°C and 30°C, infecting the root tissues and producing numerous **asexual** spores, which may be carried by **water** currents to adjacent plants. Sexual **oospores** produced further inside the root are released on decay and allow the fungus to survive in the soil several months without a host.

Preventative control (*see* hygienic growing) is important. Reliable **stock** plants should be used. **Water** supply should be checked to avoid contamination. The stock plant area should be slightly **higher** than the production area to prevent infection by drainage water. Rooting trays, compost and equipment, e.g. knives and spades, should be **sterilized** (e.g. with formalin) before use. Placing container plants on **gravel** reduces infection through the base of the pot. Some chemicals, e.g.

**etridiazole**, incorporated in compost protect the roots, but do not kill the fungus. Some species, e.g. *Juniperus horizontalis*, have some tolerance to this disease.

**Honey fungus** (*Armillaria mellea*)

This fungus primarily attacks trees and shrubs, e.g. apple, lilac and privet. In spring the foliage wilts and turns yellow. Death of the plant may take a few weeks, or several years in large trees.

The roots are infected by **rhizomorphs**, sometimes referred to as 'bootlaces', which radiate out underground from infected trees or **stumps** (Figure 11.7) for a distance of 7 m, and to a depth of 0.7 m. The nutrients they are able to conduct provide the considerable energy required for the infection of the tough, woody roots. **Mycelium**, moving up the stem to a height of several metres, is visible under the bark as white **sheets**, smelling of mushrooms. In autumn, clumps of light brown **toadstools** may be produced, often at the base of the stem. The millions of spores produced by the toadstools are not considered to be important in the infection process.

Control is difficult. Removal of the disease

**Figure 11.7** *Honey fungus on an infected stump: note the network of 'bootlaces'.*

source, the infected stump, is recommended. In large stumps, a surrounding trench in sometimes dug to a depth of 0.7 m to prevent the progress of rhizomorphs. Infected soil containing no crops may be sterilized by loosening with a fork and then applying a diluted sterilant, e.g. formalin.

**Fusarium patch on turf** (*Fusarium nivale*)

This disease appears as **irregular circular patches** of yellow then dead brown grass up to 30 cm in diameter on fine turf (Figure 11.8). Under extreme damp conditions, dead leaves become slimy and then are covered with a light-pink bloom, most evident between May and September.

Infection of the leaves by spores and hyphae occurs most seriously between 0°C and 8°C, conditions which are present under a layer of snow (hence the name **snow mould**). However, conditions of high humidity at temperatures up to 18°C may result in typical patch symptoms. Spread is by means of wind-borne asexual spores, while

**Figure 11.8** *Fusarium patch on turf: note the network of light patches.*

the fungus survives in frosty or dry summer conditions as dormant mycelium in dead leaf matter or newly infected leaves.

**Preventative** control measures (*see* soil fertility) are most important. Avoid high soil nitrogen levels in autumn, as this promotes lush, susceptible growth in autumn and winter. Avoid thatchy growth of the turf, as this encourages high humidity and thus favours the disease organism (*see* hygienic growing). Drenches of preventative fungicides, e.g. **quintozene**, applied in autumn may slow down infection of the fungus, while a summer-applied systemic fungicide, e.g. **thiophanate-methyl**, moves within the plants to achieve curative control.

**Vascular wilt diseases** (*Fusarium oxysporum* and *Verticillium dahliae*)

These two organisms infect the **xylem** tissues of horticultural plants, causing the leaves to 'flag' or wilt in hot conditions, a symptom which can also be caused by other factors, e.g. lack of soil moisture (*see* wilt), and nematode infestation (*see* root knot nematode). The wilt diseases can be recognized by yellowing and eventual browning of the lower leaves, and by brown staining of the xylem tissue when it is exposed with a knife.

Both organisms may live as **saprophytes** in the soil. *Fusarium* survives unfavourable conditions as thick-walled asexual spores, while *Verticillium* forms small **sclerotia**. Infection by both genera occurs through young roots, or after nematode attack in older roots. The fungal hyphae enter the root **xylem** tissue and then move up the stem, sometimes reaching the flowers and seeds. The diseases spread by **water-borne** asexual spores.

*Verticillium* may attack a wide range of plants, e.g. dahlia, strawberry, lilac, tomato and potato, so that rotation is not a feasible control measure. *Fusarium oxysporum*, however, exists in many distinct forms, each specializing in a different crop, e.g. tomato, broad bean or carnation. *Verticillium* more commonly attacks in spring time, having an optimum infection temperature of 20°C, while *Fusarium* is more common in summer, with an optimum temperature of 28°C.

Control is often necessary in greenhouse crops. Infected crop **residues** should be removed from the soil at the end of the growing season, as soil sterilization by **steam** or chemicals, e.g. **metham sodium**, may not penetrate to the centre of stems and roots. Peat bags may be used as a disease-free growing medium. In unsterilized soils, growers may use resistant rootstocks, e.g. in tomatoes, which are grafted onto scions of commercial cultivars.

Rotation may be employed against *Fusarium oxysporum*, as different forms attack different crops. Careful removal of infected and surrounding plants, e.g. in carnations, may slow down the progress of the diseases, especially if the soil area is drenched with a **systemic** chemical, e.g. **benomyl**, which reduces the infection in adjacent plants.

## BACTERIA

These minute organisms (Figure 11.1), which are unrelated to fungi or viruses, measure about 0.01 mm and occur as single cells which divide rapidly to build up their numbers. They are important in the conversion of soil organic matter (*see* Chapter 15), but may, in a few parasitic species, cause serious losses to horticultural plants.

### Fireblight (*Erwinia amylovora*)

This disease, which first appeared in Great Britain in 1957, can cause serious damage on members of the Rosaceae. Individual branches wilt, the leaves rapidly turning a 'burnt' chestnut brown, and when the disease reaches the main trunk it spreads to other branches and may cause death of the tree within six weeks of first infection, the general appearance resembling a burnt tree, hence the name of the disease. On slicing through an infected stem, a brown stain will often be seen. Pears, hawthorn and *Cotoneaster* are commonly attacked, while apples and *Pyracantha* suffer less commonly.

The bacterium is carried by pollinating and harmful insects (e.g. aphids), and by small droplets of rain. The spread is favoured by humid conditions and temperatures in excess of 18°C, which occur from June to September. Natural plant openings such as **stomata** and **lenticels** are common sites for infection. Flowers are the main entry point in pears. Badly infected plants produce a **bacterial slime** on the outside of the branches in humid weather. This slime is a major source of further infections.

Fireblight, once notifiable, must now be reported only in fruit-growing areas. The compulsory removal of the susceptible Laxton's Superb pear cultivar has eliminated a serious source of infection. Preventative measures such as removal of badly infected plants to prevent further infection, and replacement of hawthorn hedges close to pear orchards, help in control. Careful pruning, 60 cm below the stained wood of early infection, may save a tree from the disease. Wounds should be sealed with protective paint, and pruning implements should be sterilized with 3 per cent lysol.

### Bacterial canker (*Pseudomonas morsprunorum*)

This disease affects the plant genus *Prunus* which includes ornamental species, plum, cherry and apricot. Symptoms typically appear on the **stem** (Figure 11.9) as a swollen area exuding a light brown gum. The angle between branches is the most common site for the disease. Severe infections girdling the stems cause death of tissues above the infection, and the resulting brown foliage resembles the damage caused by **fireblight**. In May and June, leaves may become infected; dark brown leaf spots 2 mm across develop and may be blown out, giving a **'shot-hole'** effect.

The bacteria present in the cankers are mainly carried by wind-blown rain droplets, infecting leaf scars and pruning wounds in **autumn**, and young developing leaves in **summer**.

Preventative control involves the use of resistant

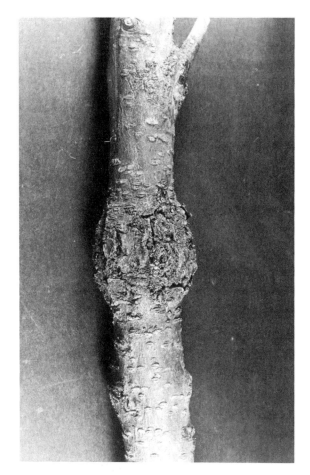

**Figure 11.9** *Bacterial canker on flowering cherry: note the swollen and cracked stem (Courtesy Plant Pathology Laboratory, MAFF).*

**rootstocks** and **scions**, e.g. in plums. The careful cutting out of infected tissue followed by application of a paint, and the use of autumn sprays of Bordeaux mixture, help reduce this disease.

### Soft rot (*Erwinia carotovora*)

This bacterium affects stored potatoes, carrots, bulbs and iris, where the bacterium's ability to dissolve the cell walls of the plant results in a mushy **soft rot**. High temperatures and humidity caused by poor ventilation promote infection through lenticels, and major losses may occur. A related strain of this bacterium causes **black leg** on potatoes in the field.

**Preventative** control measures are important. Crops should be damaged as little as possible when harvesting, and diseased or damaged specimens should be removed before storage. Hot, humid conditions should be avoided in store. No curative measures are available.

## VIRUSES

### Structure and biology

**Viruses are extremely small; much smaller even than bacteria (Figure 11.1). They appear as rods or spheres when seen under an electron microscope. The virus particle is composed of a nucleic acid core surrounded by a protective protein coat. The virus, on entering a plant cell, takes over the organization of the cell nucleus in order to produce many more virus particles. Since the virus itself lacks any cytoplasm cell contents, it is sometimes considered to be a non-living unit. The virus's close association with the plant cell nucleus presents difficulties in the production of a curative virus control chemical that does not also kill the plant. No such commercial 'viricide' has yet been produced.**

In recent years the broad area called 'virus diseases' has been closely investigated. Virus particles have, in most cases, been isolated as the cause of disease, e.g. cucumber mosaic. Other **agents of disease** to be discovered are **viroids** (e.g. in chrysanthemum stunt disease), and these are much smaller than viruses. **Mycoplasmas** (e.g. aster yellows disease) are larger than and unrelated to viruses. In some diseases, e.g. 'big vein' on lettuce and 'reversion disease' on blackcurrant, a causitive agent has yet to be isolated. All agents are placed together in this section.

A number of organisms (**vectors**) transmit viruses from plant to plant. **Peach potato aphid** is capable of transmitting over 200 strains of virus (e.g. cucumber mosaic) to different plant species; the aphid stylet injects salivary juices containing virus into the **parenchyma** and **phloem** tissues, and, along the phloem, the virus may travel to other parts of the plant. Other vector/virus combinations include blackcurrant big bud mite and reversion virus; bean weevils and broad bean stain virus; migratory soil nematodes, *Xiphinema* and arabis mosaic; and *Olpidium* soil fungus and big vein agent on lettuce. Other important methods of transmission involve **vegetative** material (e.g. chrysanthemum stunt viroid and plum pox), infected **seed** (e.g. bean common mosaic virus), **seed testa** (e.g. tomato mosaic virus), **mechanical transmission** by hand (e.g. tomato mosaic virus).

### Symptoms

The presence of a damaging virus in a plant is recognized by the horticulturist only by the symptoms, although he may consult the virologist, whose identification techniques include electron microscopy, transmission tests on sensitive plants, e.g. *Chenopodium* spp., and serological reactions against specific antisera.

**Leaf mosaic**, a yellow mottling, is the most common symptom (e.g. tomato mosaic virus). Other symptoms include leaf **distortion** into feathery shapes (cucumber mosaic virus), flower **colour** streaks (e.g. tulip break virus), fruit **blemishing** (tomato mosaic, plum pox), internal **discoloration** of tubers (tobacco rattle virus causing 'spraing' in potatoes), and **stunting** of plants (chrysanthemum stunt viroid). Symptoms similar to those described above may be caused by misused herbicide sprays, genetic '**sports**', poor soil fertility and structure (*see* deficiency symptoms) and mite damage.

In the following descriptions of major viruses, Latin names of genus and species are omitted, since no consistent classification is yet accepted.

### Cucumber mosaic

Several strains of virus cause this disease, and many families of plants are attacked. On cucum-

bers, a mottling of young leaves occurs (Figure 11.10), followed by a twisting and curling of the whole foliage, and fruit may show yellow sunken areas. On the shrub *Daphne oderata*, a yellowing and slight mottle is commonly seen on infected foliage, while *Euonymus* foliage produces bright yellow leaf spots. Infected tomato leaves are reduced in size (**fern-leaf** symptom).

The virus may be transmitted by infected hands, but more commonly an aphid (e.g. peach-potato aphid) is involved. Many crops (e.g lettuce, maize, *Pelargonium*, privet) and weeds (e.g. fat hen and teasel) may act as a reservoir for the virus.

Since there are no curative methods for control, care must be taken to carry out **preventative** methods. Choice of **uninfected stock** is vital in vegetatively propagated plants, e.g. *Pelargonium*. Careful control of **aphid vectors** may be important where susceptible crops (e.g. lettuce and cucum-

**Figure 11.10** *Cucumber mosaic: note the irregular blotches over the leaf surface (Courtesy Rothamsted Experimental Station).*

bers) are grown in succession. Removal of infected **weeds**, particularly from greenhouses, may prevent widespread infection.

### Tulip break

The petals of infected tulips produce irregular coloured streaks and may appear distorted. Leaves may become light green, and the plants stunted after several years infection.

The virus is transmitted mechanically by knives, while three aphid vectors are known; the bulb aphid in stores, the melon aphid in greenhouses, and the peach-potato aphid outdoors and in greenhouses.

Preventative control must be used against this disease. Removal of **infected** plants in the field prevents a source of virus for aphid transmission. **Aphid** control in field, store and greenhouse further reduces the virus's spread.

### Tomato mosaic

This disease may cause serious losses in tomatoes. Infected seedlings have a stunted, spiky appearance. On more mature plants leaves have a pale green **mottled** appearance (Figure 11.11), or some-times a bright yellow ('aucuba') symptom. The stem may show brown **streaks** in summer when growing conditions are poor, a condition often resulting in death of the plant. Fruit yield and quality may be lowered, the green fruit appearing **bronze**, and the ripe fruit hard, making the crop unsaleable.

The virus may survive within the **seed coat** (testa) or endosperm. **Heat** treatment of **dry** seed at 70°C for four days by seed merchants helps remove initial infection. Infected **debris**, particularly roots, in the soil enables the virus to survive from crop to crop, and soil temperatures of 90°C for ten minutes are normally required to kill the organism. Peat growing bag and nutrient-film methods enable the grower to avoid this source of infection. The virus is spread very easily by sap. Hands and tools should be washed in soapy water after working with infected plants. Clothing may harbour the virus. The period from plant infection to symptom expression is about fifteen days.

Cultivars and rootstocks containing several factors for **resistance** are commonly grown, but changing virus strains may overcome this resistance. A **mild strain** spray inoculation method has been used at the seedling stage to protect non-resistant cultivars from infection with severe strains. Extreme care is required to avoid mosaic-contaminated equipment when using this method.

**Figure 11.11** *Tomato mosaic: note the fine mosaic.*

### Plum pox

This disease, sometimes called 'Sharka', has increased in importance in Great Britain since 1970, after its introduction from mainland Europe. Plums, damsons, peaches, blackthorn and orna-mental plum are affected, while cherries and flowering cherries are immune. Leaf symptoms of **feint interveinal yellow blotches** can best be seen on leaves from the centre of the infected tree. The most **reliable** symptoms, however, are found on fruit (Figure 11.12), where sunken **dark blotches** are seen. Ripening of infected fruit may be several weeks premature, yield losses may reach 25%, and the fruit is often **sour**.

**Figure 11.12** *Plum pox: note the light patches on the fruit (Courtesy Plant Pathology Laboratory, MAFF).*

The virus enters a nursery or orchard through infected **planting stock**. Spread of the virus during the early summer and autumn by **aphids** (e.g. leaf curling plum aphid) is **slow** in widely spaced orchards, but more rapid in closely planted nurseries, e.g. nursery stock areas.

**Preventative** control is the only option open to growers. Clean Ministry-certified stock should be used. Routine aphid-controlling **insecticides** should be applied in late spring, summer and autumn. Suspected infected tress must be reported to the Ministry of Agriculture (*see* notifiable diseases). Infected trees must be **removed** and **burnt**.

### Chrysanthemum stunt viroid

This disease, found only on plants of the Compositae family and mainly on the *Chrysanthemum* genus, produces a stunted plant, often only half normal size but **without** any distortion. Flowers often open one week earlier than normal, and may be small and lacking in colour.

The virus enters gardens and nurseries through infected **cuttings**, and is readily transmitted by leaf contact and by **handling**. Symptoms may take several months to appear, thus seriously reducing the chance of early removal of the disease source.

**Preventative** control must be used by the grower. Certified planting material derived from **heat-treated meristem** stock (*see* tissue culture) reduces the risk of this disease.

### Arabis mosaic

This infects a **wide** range of horticultural crops. On strawberries, yellow spots or mottling are produced on the leaves, and certain cultivars become **severely stunted**. On ornamental plants, e.g. *Daphne odorata*, yellow rings and lines are seen on infected leaves, and the plants may slowly die back, particularly when this virus is associated with cucumber mosaic inside the plant. Several weeds, e.g. chickweed and grass spp., may harbour this disease, and in soft fruit severe attacks of the disease may occur when planted into ploughed-up grassland.

The virus is transmitted by a common soil-inhabiting nematode, *Xiphinema diversicaudatum*, which may retain the virus in its body for several months.

Control of this disease may be achieved only by **preventative** methods. Virus-free soft fruit **planting** material is available through the Nuclear Stock Scheme. **Fumigation** of soil with chemicals such as **dichloropropene**, applied well before planting time, eliminates many of the eelworm vectors. No curative chemical is available to eliminate the virus inside the plant.

### Reversion disease on blackcurrants

This virus disease seriously reduces blackcurrant yields. Flower buds on infected bushes are almost hairless on close inspection, and appear brighter in colour than healthy buds. Infected leaves may have fewer main veins than healthy ones. After several years of infection, the bush may cease to produce fruit, the effect of different virus strains producing different levels of sterility.

The virus is transmitted by the **blackcurrant gall mite**, and reversion infected plants are particularly susceptible to attack by this pest.

Removal and burning of infected plants is an important form of control. Use of **certified plant material**, raised in areas away from infection and vectors, is strongly recommended. Control of the mite **vector** in spring and early summer has already been described.

## PRACTICAL EXERCISES

1 Dig up the soil of a 'fairy-ring' on turf. Determine how deep the white mycelium penetrates.
2 Place a large mushroom over a sheet of white paper. Leave overnight and observe the spore trace produced.
3 Observe the range of toadstools present in a woodland, and examine rotting wood for evidence of fungus mycelium. Use a fungus 'flora' for identification.
4 Seal a slice of fresh bread, or an orange, in a polythene bag. Determine the period, in days, taken for the blue spore-production stage of the blue mould to occur.
5 Cut open a soft, rotten potato, or remove the stems of flowers left in a vase with stale water. Notice the characteristic slimy bacterial growth.

## FURTHER READING

British Crop Protection Council. *The UK Pesticide Guide* (1993).

Buczacki, S., and Harris, K. *Guide to Pests, Diseases and Disorders of Garden Plants* (Collins, 1992).

Cooper, J.I. *Virus Diseases of Trees and Shrubs* (Institute of Terrestrial Ecology, 1979).

Dahl, M.H., and Thygesen, T.B. *Garden Pests and Diseases* (Blandford Press, 1974).

Fletcher, J.T. *Diseases of Greenhouse Plants* (Longman, 1984).

Hope, F. *Turf Culture: A Manual for Groundsmen* (Cassell, 1990).

Morgan, W.M., and Ledieu, M.S. *Pest and Disease Control in Glasshouse Crops* (British Crop Protection Council, 1979).

Pirone, P.P. *Diseases and Pests of Ornamental Plants* (Wiley, 1985).

Scopes, N., and Stables, L. *Pest and Disease Control Handbook* (British Crop Protection Council, 1989).

Snowdon, A.L. *Post-harvest Diseases and Disorders of Fruits and Vegetables* (Wolfe Scientific, 1991).

Toms, A.M., and Dahl, M.H. *Pests and Diseases of Fruit and Vegetables* (Blandford Press, 1976).

*Welcome to the World of Environmental Products* (Rhone-Poulenc, 1992).

Wheeler, B.E.J. *An Introduction to Plant Diseases* (Wiley, 1969).

# 12

# Control Measures

*In the four preceding chapters, import-
ant weeds, pests, and diseases have been
described with an emphasis on symptoms
and damage, life cycles, and brief comments
on control relevant to the particular causitive
organism. In this chapter, the main types of
control measures are described in some detail
to include legal aspects, crop management
methods, use of beneficial organisms, and
the rather complex area of chemical control.
The horticulturist should aim to use as many
appropriate methods as possible within a crop
cycle to bring about precise and efficient control. Distinction is drawn between* **preventative** *and* **curative** *control methods. The
concept of* **economic damage** *is discussed in
relation to supervised control.*

## LEGISLATION AND CONTROL

Before 1877, no legal measures were available in
the UK to prevent importation of plants infested
with pests such as Colorado beetle. Measures
taken at ports from that year onwards were
brought together in the 1927 Destructive Insects
and Pests Acts, empowering government officials
to inspect and, if necessary, refuse plant imports.
Within this Act, the 'Sale of Diseased Plants
Order' placed on the grower the responsibility for
recognition and reporting of serious pests and
diseases, e.g. blackcurrant gall mite and silver
leaf of plums. Lack of education and enforcement
led to the need for specific orders relevant to
particular current problems, e.g. in 1958 the
fireblight-susceptible pear cultivar, Laxton's
Superb, was declared **notifiable** and prohibited
under the Fireblight Disease Order. Recent orders
have helped prevent outbreaks of white rust
on chrysanthemums, plum pox virus and two
American leaf miner species; less success has
been achieved with the western flower thrip organism. Further importation legislation under the
'1967 Plant Health Act' prohibits the landing of
any non-indigenous pest or disease by aircraft or
post, and the 'Importation of Plants, Produce and
Potatoes Order 1971' specifically names prohibited crops, e.g. plum rootstocks from eastern
Europe, crops subject to a healthy-plant (**phyto-
sanitary**) certificate, e.g. potato tubers from
Europe, and crops to be examined, e.g. *Acacia*
shrubs and apricot seeds. The operation of these
orders is supervised by the Plant Health Branch
of the Ministry of Agriculture. Complete success
in preventing the introduction of damaging organisms may be limited by dishonest importations
and by the difficulty of detection of some diseases,
especially viruses.

The Weeds Act 1959 places a legal obligation
on each grower to prevent the spread of creeping
thistle, spear thistle, curled dock, broad-leaved
dock and ragwort.

# CULTURAL CONTROL

Horticulturists, in their everyday activities, may remove or reduce damaging organisms, and thus **protect** the crop. Below are described some of the more important methods used.

## Cultivation

**Ploughing** and **rotavating** of soils enable a physical improvement in soil structure as a preparation for the growing of crops. The improved drainage and tilth may reduce damping-off diseases, disturb annual and perennial weeds, e.g. chickweed and couch grass, and expose soil pests, e.g. leatherjackets and cutworms, to the eager beaks of birds. Repeated rotavation may be necessary to deplete the reserves of perennial weeds, e.g. couch grass. **Hoeing** annual weeds is an effective method, provided the roots are fully exposed, and the soil dry enough to prevent their re-establishment.

## Partial soil sterilization

Greenhouse soils are commonly sterilized by high pressure steam released to penetrate downwards into the soil, which is covered by heat resistant plastic sheeting (**sheet steaming**). The steam condenses on contact with soil particles, and moves deeper only when that layer of soil has reached steam temperature. Some active soil pests, e.g. symphilids, may move downwards ahead of the steam 'front'. The temperatures required to kill most nematodes, insects, weed seeds and fungi are 45°C, 55°C, 55°C and 60°C respectively. Beneficial bacteria are not killed below 82°C, and therefore growers attempt to reach but not exceed this soil temperature. In this way, organisms difficult to sterilize, such as fungal sclerotia, and *Meloidogyne* and *Verticillium* in root debris may be killed. Sheet steaming is effective only to depths of about 15 cm, and its effect is reduced when soil aggregates are large and hard to penetrate, or when soils are wet and hard to heat up.

When soil pests and diseases occur deep in the soil, heating pipes may be placed below the soil surface, as grids or spikes, to achieve a more thorough effect. The 'steam-plough' achieves a similar result, as it is winched along the greenhouse. If soil is to be used in composts it should be sterilized (*see* sterilizing equipment). The clear advantage of soil sterilization may occasionally be lost if a serious soil fungus (e.g. *Pythium*) is accidentally introduced into a crop where it may quickly spread in the absence of fungal competition.

## Chemical sterilization

This involves the use of substances toxic to most living organisms, thus necessitating application during the fallow, or intercrop, period. **Soil applied** chemicals include methyl bromide (applied as a gas), metham sodium (commonly applied through a nutrient diluter), dazomet (applied as a granule), and dichloropropene (applied by an injection apparatus against nematodes). The fumigant action of these substances is prolonged by rolling the soil, or covering it with plastic sheeting (in the case of the methyl bromide). Care should be taken that no chemical residues are allowed to remain, and thus kill off early stages of the succeeding crop. **Greenhouse structures** may be sterilized by toxic gases and liquids such as formaldehyde, formic acid and burning sulphur. Common pests and diseases such as whitefly, red spider mite and grey mould may be greatly reduced by this intercrop method of control.

## Soil fertility

While the correct content and balance of major and minor nutrients (*see* Chapter 16) in the soil are recognized as vitally important for optimum crop yield and quality, excessive nitrogen levels may encourage the increase of insects such as peach potato aphid, fungi, e.g. grey mould, and bacteria, e.g. fireblight. Adequate levels of potassium help control fungal diseases, e.g. *Fusarium* wilt on carnation, and tomato mosaic virus. Club root disease of brassicas is less damaging in soil with a pH greater than 6, and lime may be incorporated before planting these crops to achieve this aim. Amenity horticulturists apply mulches, e.g. composted bark, grass cuttings or straw, to bare soil in order to control annual weeds by excluding the source of light. Black polythene sheeting is used in soft fruit production to achieve a similar objective.

### Crop rotation

Important pests and diseases, such as cyst nematode and club root, attack specific crops (potato and cruciferous plants respectively). They are slow in their dispersal and, being soil borne, are difficult to control. By the simple method of planting a given crop in a different plot each season, such pests or diseases are excluded from their preferred host crop for several seasons. This method does not work well against unspecific problems, for example grey mould, or rapidly spreading organisms such as aphids.

### Planting and harvesting times

Some pests emerge from their overwintering stage at about the same time each year, e.g. cabbage root fly in late April. By planting early to establish tolerant brassica plants before the pest emerges, a useful supplement to chemical control is achieved. The deliberate planting of early potato cultivars enables harvesting before the maturation of potato cyst nematode cysts, so that damage to the crop and the release of the nematode eggs is avoided. Annual weeds may be induced to germinate in a prepared seedbed by irrigation. After they have been controlled with a contact herbicide, e.g. paraquat, a crop may then be sown into the undisturbed bed or **stale seedbed**, with less chance of further weed germination.

### Clean seed and planting material

Seed producers take stringent precautions to exclude weed seed contaminants and pests and diseases from their seed stocks. While weed seeds are, in the main, removed by mechanical separators, and insects can be killed by seed dressings, systemic fungal seed infections, e.g. celery leaf spot disease, are best controlled by immersion of dry seed in a 0.2% thiram solution at 30°C for twenty-four hours (**thiram soak treatment**). The seed is then re-dried. Equal care is taken to monitor seed crops likely to carry virus disease (e.g. tomato mosaic). Curative control by **dry heat** at 70°C for four days is usually effective, although it may reduce subsequent seed germination rates. **Vegetative propagation** material is used in all areas of horticulture, as bulbs (e.g. tulips and onions), tubers (e.g. dahlias and potatoes), runners (e.g. strawberries), cuttings (e.g. chrysanthemums and many trees and shrubs), and graft scions in trees. The increase of nematodes, viruses, fungi and bacteria by vegatative propagation is a particular problem, since the organisms are inside the plant tissues, and since the plant tissues are sensitive to any drastic control measures. **Inspection** of introduced material may greatly reduce the risk of this problem. Soft narcissus bulbs, chrysanthemum cuttings with an internal rot, whitefly or red spider mite on stock plants, virus on nursery stock, are all symptoms which would suggest either careful sorting, or rejection of the stocks. Accurate and rapid methods of **virus testing** (using test plants, electron microscopy, and staining by ELISA technique) now enable growers quickly to learn the quality of their planting stocks. Fungal levels in cuttings (e.g. *Fusarium* wilt of carnations) can be routinely checked by placing plant segments in sterile nutrient culture.

**Warm water-treatment** is used for pests such as stem and bulb nematodes in narcissus bulbs. Immersion of bulbs for two hours at 44°C controls the pest without seriously affecting bulb tissues. Chrysanthemum stools and strawberry runners may similarly be treated, using temperature/time combinations favourable to each crop. **Viruses** (e.g. aspermy virus on chrysanthemum) are more difficult to control, since they are more intimately associated with the plant cells. Virus concentrations may be greatly reduced in meristems of stock plants grown at temperatures of **40°C for about a month**. This has enabled the production of **tissue-cultured** disease-free stock material of both edible and non-edible crops.

### Hygienic growing

During the crop, the grower should aim to provide optimum conditions for growth. Water content of soil should be adequate for growth (*see* field capacity), but not be so excessive that root diseases (e.g. damping off in pot plants, club root of cabbage, and brown root rot of conifers) are actively encouraged. Water sources should be analysed for *Pythium* and *Phytophthora* species if damping-off diseases are a constant problem. Covering, and regular cleaning, of water tanks to prevent the breeding of these fungi in rotting organic matter may be important in their control. Conifer nursery stock grown on raised gravel beds are less likely to suffer the water-borne spread of conifer brown rot. Carnations are grown in **isolated beds** or **peat modules** to reduce spread of wilt-inducing organisms (e.g. *Fusarium* spp.).

**High humidity** encourages many diseases. In greenhouses, the careful timing of daily overhead irrigation, and of ventilation (to reduce overnight condensation on leaves or flowers), may greatly reduce levels of diseases such as grey mould on pot plants, or downy mildew on lettuce.

**Transmission** of pests and diseases from plant to plant or field to field can be slowed down. Virus diseases such as tomato mosaic virus may be confined by delaying till last the deleafing or harvesting or infected plants. Washing knives and hands regularly in warm, soapy water will reduce subsequent viral spread. Soil-borne problems, e.g. club root, eelworms and damping-off diseases, are easily carried by boots and tractor wheels. Foot and wheel **dips**, containing a general chemi-

cal sterilant, e.g. formaldehyde, have successfully been used, especially in preventing damping-off problems in greenhouses.

### Traps and repellants

Measures described in this section aim to prevent the pest or disease from arriving at the crop. A **grease band** wrapped around an apple trunk prevents the wingless female winter moth's attempted crawl up the tree to lay eggs in winter. A fine mesh screen placed in front of the ventilator fan prevents the entry of many mushrooms and glasshouse pests, e.g. sciarid flies. A **trench** dug around a large stump infected with honey fungus may prevent the '**bootlaces**' from radiating out to infect other plants. Interplanting of onions amongst carrots may deter the carrot fly from attacking its host crop. **Pheremone traps** containing a specific synthetic attractant are used in apple orchards to lure male codling and tortrix moths onto a sticky surface, thus enabling an accurate assessment of their numbers, and a more effective control. Repellant **colophony** chemical, impregnated in sacking, is used to deter rabbits from eating amenity border plants.

### Alternate hosts

Alternate hosts harbouring pests and diseases should be removed where possible. Soil-borne problems, e.g. club root of cabbage, and free-living eelworms on strawberries, are harboured by shepherd's purse (*Capsella bursapastoris*) and chickweed (*Stellaria media*) respectively. Groundsel (*Senecio vulgaris*) is an alternate host of the air-borne powdery mildew of cucumber, while docks (*Rumex* spp.) act as a reservoir of dock sawfly, which damages young apple trees.

### Removal of infected plant material

With **rapid-increase** problems, e.g. peach potato

aphid, and white rust of chrysanthemum fungus in greenhouses, removal is practicable in the early stages of the crop, but becomes unmanageable after the pest or disease has increased in number and dispersed throughout the plants. **Slow-increase** problems, e.g. *Fusarium* wilt disease on carnations, and vine weevil larvae, may be removed throughout the crop cycle, but the infected roots and soil must be carefully placed in a bag to prevent dispersal of the problem. In outdoor production, labour costs often prevent removal during the growing season. However, chemical destruction of blight-infected potato foliage with a herbicide, e.g. diquat, reduces infection of the tubers. **Burning** of post-harvest leaf material and lifting of root debris after harvest (against grey mould on strawberries, and club root on brassicas respectively) may help prevent problems in the next crop. In tree species, routine **pruning** operations may remove serious pests, such as fruit tree red spider mite, and diseases, such as powdery mildew. Tree stumps harbouring serious underground diseases, e.g. honey fungus, should be removed where practicable.

## RESISTANCE

Wild plants show high levels of resistance to most pests and diseases. In the search for high yields and extremes of flower shape and colour, plant breeders have often failed to include this wild plant resistance. However, in crops such as antirrhinum, lettuce and tomatoes, one or more resistance **genes** have been deliberately incorporated to give protection against rust, downy mildew and tomato mosaic virus respectively. The disease organisms may **overcome** these genetic barriers, and the crop thus again becomes infected. Growers may sow a sequence of cultivars (e.g. lettuce), each with different resistance genes, in order that the disease organism (e.g. downy mildew) is constantly exposed to a new resistance barrier, and thus limit the disease. **Vegetatively** propagated species (e.g. potatoes) and **tree** crops (e.g. apple) which remain genetically unaltered

for many years, are now being bred with high levels of 'wild plant' resistance (to blight and powdery mildew respectively on potato and apple) so that the crops may more permanently resist these serious problems. Crop resistance to **insects** is now being more seriously considered by plant breeders. Some lettuce cultivars are resistant to lettuce root aphid (*Pemphigus bursarius*).

## BIOLOGICAL CONTROL

Many pests of outdoor horticultural crops, e.g. peach potato aphid, are **indigenous** (i.e. they originally evolved, and are still present, in wild plant communities of this country. Pest populations are often reduced in nature by other organisms which, as **predators**, eat the pest, or as **parasites**, lay eggs within the pest. These beneficial organisms, found also on horticultural crops, are to be encouraged, and in some cases deliberately introduced.

The **black-kneed capsid** (*Blapharidopterus angulatus*) found on fruit trees alongside its pestilent relative, the **common green capsid**, eats more than a thousand fruit tree red spider mites per year. Its eggs are laid in August and survive the winter. Winter washes used against apple pests and diseases often kill off this useful insect. The closely related **anthocorid** bugs, e.g. *Anthocoris nemorum*, are predators on a wide range of pests, e.g. aphids, thrips, caterpillars, and mites, and have recently been used in controlled introduction with greenhouses.

Some species of **lacewings**, e.g. *Chrysopa carnea*, lay several hundred eggs per year on the end of fine stalks, located on leaves. The resulting hairy larvae predate on aphids and fruit tree red spider mite, reaching their prey in leaf folds which ladybirds cannot reach.

The forty species of **ladybird** beetle are a welcome sight to the professional horticulturist and layman alike. Almost all are predatory. The red 'two-spot ladybird' (*Adalia bipunctata*) emerges from the soil in spring, mates and lays about a thousand elongated yellow eggs on the leaves of a range of weeds and crops, e.g. nettles and beans, throughout the growing season. The emerging slate-grey and yellow larvae and the adults feed on a range of aphid species. The 'ground beetle' (*Bembidion lampros*) actively predates on soil pests, e.g. cabbage root fly eggs, greatly reducing their numbers.

**Hoverflies** superficially resembling wasps are commonly seen darting above flowers in summer. Several of the 250 British species, e.g. *Syrphus ribesii*, lay eggs in the midst of aphid colonies, and the legless larva consumes large numbers of aphids. Predatory mites, e.g. *Typhlodromus pyri* on fruit tree red spider mite, may contribute importantly to pest control. The numerous species of web-forming and hunting **spiders** help in a largely unspecific way in the reduction of all forms of insects.

About 3000 **wasp** spp. of the families Ichneumonidae, Braconidae and Chalcidae are parasitic on insects in Britain. The red ichneumon (*Opion* spp.) lays eggs in many moth caterpillars, The cabbage white caterpillar may bear 150 parasitic larvae of the braconid wasp (*Apanteles glomeratus*) which pupate outside the pest's body as yellow cocoon masses. The woolly aphid on apples may be reduced by the chalcid *Aphelinus mali*.

The **spiracles** of insects provide access to specialized parasitic fungi, particularly under damp conditions. Aphid numbers may be quickly reduced by the infection of the fungus *Entomophthora aphidis*, while codling moth caterpillars on apple may be enveloped by *Beauveria bassiana*. Cabbage white caterpillar populations are occassionally much reduced by a virus which causes them to burst.

The predators and parasites previously described may be irregular in their distribution, and time of emergence. For reliable results, deliberate introduction may be necessary. The raising and measured application of certain predators and parasites are now common features of greenhouse plant protection, particularly where pests have developed **resistance** to chemicals. The greenhouse represents a controlled environment where pest−predator interactions can be accurately predicted. Developments in outdoor plant production may yield similarly successful results in future. Some of the more effective species being used are described below.

**Phytoseiulus persimilis** (Figure 12.1)

This is a globular, deep orange, predatory tropical mite used in greenhouse production to control **glasshouse red spider mite** (*see* page 83). It is raised on spider mite-infected beans, and then evenly distributed throughout the crop, e.g. cucumbers at the rate of one predator per plant. Some growers who suffer repeatedly from the pest first introduce the red spider mite throughout the crop at the rate of about five mites per plant a week before predator application, thus maintaining even levels of pest−predator interaction. The predator's short egg−adult development period (seven days), laying potential (fifty eggs per life cycle), and appetite (five pest adults eaten per day), explain its extremely efficient action.

**Encarsia formosa** (Figure 12.2)

This is a small (6 mm) wasp, which lays an egg into the glasshouse **whitefly** (*see* page 72) scale, causing it to turn black and eventually release

**Figure 12.1** *Phytoseiulus predator: note the difference in appearance between the round, long-legged predators (left) and the glasshouse red spider mite (right) (Courtesy Glasshouse Crops Research Institute).*

**Figure 12.2** *Encarsia parasite of glasshouse whitefly (Courtesy Glasshouse Crops Research Institute).*

another wasp. This parasite is raised on whitefly-infested tobacco plants. It is introduced to the crop, e.g. tomato, at a rate of 100 blackened scales per 100 plants. The parasite's introduction to the crop is successful only when there is less than one whitefly per ten plants, and its mobility (about 5 m) and successful parasitism are effective only at temperatures greater than 22°C when its egg-laying ability exceeds that of the whitefly. The wasp lays most of its sixty or more eggs within a few days of emergence from the black scale. Thus a series of weekly or fortnightly applications from late February onwards ensures that parasite egg-laying covers the susceptible whitefly scale stage.

**Other biological control organisms**

Peach-potato aphid is parasitized by the wasp,

*Aphidius matricariae*, and fungus *Verticillium lecanii*. Chrysanthemum leaf miner is controlled by the parasitic wasps, *Opius* and *Diglyphus* spp. Mealy bug is rapidly eaten by the tropical green and brown ladybird *Cryptolaemus montrouzieri*. Caterpillars may be killed by an extract of the bacterium *Bacillus thuringiensis*. Vine weevil larvae can be infected and killed by the parasitic nematode, *Steinernema bibionis*. Early infection, by the silver leaf fungus, on ornamental *Malus* and fruit trees is reduced by the inoculation of purified cultures of the green soil fungus *Trichoderma viride*. A *mild form* of tomato mosaic virus inoculated into the young tomato seedlings enables the plants to develop resistance to the severe virus strain.

A combination of biological methods may be used on some crops, e.g. chrysanthemums, tomatoes and cucumbers, in order to control a range of organisms (*see* integrated control). Biological control programmes enable pollution-free elimination of some major pests without the development of **resistant** pest organisms. An understanding of the pests' and biological control organisms' life cycles is, however, necessary to ensure success. Several specialist firms now have contracts to apply biological control organisms to greenhouse units.

The careful selection of pesticides (*see* pirimicarb) enables simultaneous biological and chemical control, to the grower's advantage.

**CHEMICAL CONTROL**

This method of control aims ideally to use a selected chemical for reduction of weed, pest or disease without harming man, crop or wild life. This aim is not always achieved, although increasingly stringent demands are placed on the chemical manufacturer, both in terms of the chemical's efficiency and its safety. In past centuries, pests, e.g. apple woolly aphid, were sprayed with natural products, e.g. turpentine and soap, while weeds were removed by hand. In the nineteenth century, the chance development of **Bordeaux mixture** from inorganic copper sulphate and slaked lime, and in the early twentieth century the expansion of the organic chemical industry, enabled a change of emphasis in crop protection from cultural to chemical control. **The word 'pesticide' is used in this book to cover all crop protection chemicals which include herbicides (for weeds), insecticides (for insects), acaricides (for mites), nematicides (for nematodes), and fungicides (for fungi).** About three million tonnes of crop protection chemicals are used worldwide each year, two fifths being herbicides, two fifths insecticides, and one fifth fungicides.

**Active ingredient**. Each container of commercial pesticide contains several ingredients. The active ingredient's role is to kill the weed, pest or disease. More detailed lists of the range of active ingredients can be found in government literature. The other constituents of pesticides are described under **formulation**.

### Herbicides

Herbicides which are applied to the seed bed or growing crop must kill the weed but leave the crop undamaged (**selective action**).

This selective action may succeed for one of several reasons. Different plant **families** often show such differences. The broad-leaved turf weed, daisy (*Bellis perennis*, a member of the Compositae) is controlled by 2,4-D, leaving the turf grasses (Graminae) unaffected.

Secondly, the correct **concentration** of selective chemicals may be vital if the crop is to remain untouched. The following relative values (ppm) for the amount of propyzamide herbicide required to kill different plant species illustrate this point: **crops**—carrot 0.8, cabbage 1.0, lettuce 78.0; **weeds**—knotgrass 0.08, black nightshade 0.2, fat hen 0.2, pennycress 0.6, and groundsel 78.0. It can thus be seen that a concentration of 25 parts per million of propyzamide applied to lettuce would leave the crop unaffected, but control all the weeds except groundsel. A low concentration of simazine is used to control annual weeds around raspberry plants, while a high concentration gives

total control on paths and other uncultivated areas.

A third form of selectivity operates by **correct timing** of herbicide application. A seedbed with crop seeds below and weed seeds germinating at the surface may receive a **contact** chemical, e.g. paraquat, which permits germination of the crop without weed competition. A similar effect is achieved when a **residual** herbicide, e.g. propachlor, is sprayed on to the soil surface to await weed seed germination. The situations for weed control are summarized in Figure 12.3.

Herbicides may conveniently be divided into two main groups, the foliage-acting, and the soil-acting (residual) chemicals.

**Foliage-acting** herbicides enter the leaf through

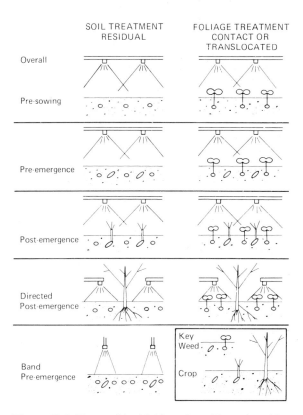

**Figure 12.3** *Types of herbicide action. (Reproduced by permission of Blackwell Scientific Publications).*

fine pores in the cuticle, or the stomata. The herbicide may move through the **vascular** system (**translocated** chemicals) to all parts of the plant before killing plant cells, or it may kill on **contact** with the leaf. Four active ingredients are described, each belonging to a different chemical group, and each having a different effect on weeds.

**Paraquat** is commonly used to scorch and kill top growth of a wide spectrum of weeds in stale seedbeds, after harvest, in perennial crops, or in wasteland. Although translocated when in dilute concentrations, its rapid, light-induced unselective contact action is most commonly utilized. It is quickly absorbed, in damp soils, by **clay** particles, thus allowing planting soon after its application. Its absorption prevents any problem of residual action except in extremely dry summers. It should **never** be used on foliage of growing plants, although carefully directed sprays between the rows of growing soft fruit are recommended.

**Amitrole** is used in similar situations to paraquat, but is more residual, surviving in the soil for several weeks. It stops photosynthesis, scorching both grass and broad-leaved weeds (**unselective**). It is specially useful on uncropped land and, when applied in autumn, it is translocated to underground rhizomes of **couch** which are then killed. It should not be sprayed on to the foliage of growing plants.

**2,4-D** is an **auxin** and causes uncontrolled abnormal growth on leaves, stems and roots of **broad-leaved** weeds, which eventually die. It is a useful selective herbicide on turf because the protected meristems of grasses can survive unaffected. It must be kept well away from nearby border plants and from some crops, e.g. tomatoes, which are extremely sensitive to minute quantities.

**Glyphosate** enters the foliage of actively growing annual and perennial weeds (**unselective**) and is **translocated** (*see* page 27) to underground organs, subsequently killing them. It is commonly used several weeks before drilling or planting of crops; around perennial plants such as apples; or in established nursery stock tress. Glyphosate is inactivated in soils (particularly **peats**), thus pre-

venting damage to newly sown crops. It may cause damage if spray-drift to adjoining plants or fields occurs.

**Soil-acting herbicides** are either sprayed on to the soil surface (*see* Figure 12.3) or soil incorporated. They must be persistent (**residual**) for several weeks or months to kill the seedling before or after it emerges. Root hairs are the main point of entry. Increased rates may be necessary for **peat soils** which inactivate some herbicides. The chemical may be applied as a spray or granule before the crop is sown (**pre-sowing** stage), before the crop emerges (**pre-emergence**) or, with more selective chemicals, after the crop emerges (**post-emergence**). Three active ingredients are described, each belonging to a different chemical group, and each having a different effect on weeds.

**Chlorpropham**, a relatively insoluble compound, is applied as a pre-emergent spray to control many germinating weeds species, e.g. chickweed, in crops such as bulbs, onions, carrots and lettuce. It usually persists for less than three months in the soil. In light, porous soils with low organic matter, its rapid penetration to underlying seeds make it an unsuitable chemical. Earth-worm numbers may be reduced by its presence.

**Propachlor**, a relatively insoluble compound, is applied as a pre-sowing or pre-emergent spray to control a wide variety of annual weeds in brassicas, strawberries, onions and leeks. For weeds in established herbaceous borders (e.g. rose), the granular formulation gives a residual protection against most germinating broad-leaved and grass weeds.

**Dichlobenil** gives **total** control against germinating weeds, couch grass and some perennial weeds in waste ground, soft fruit, top fruit and established ornamental shrub and tree areas. The compound remains in the soil for more than six months, and young crops should not be planted within a year of its application.

**Mixtures.** The horticulturist must deal with a wide range of annual and perennial weeds. The somewhat specialized action of some of the herbicide active ingredients previously described may be inadequate for the control of a broad weed spectrum. In the case of chlorpropham, for example, the addition of diuron enables an improved control of charlock and groundsel, while a different formulation containing chlorpropham plus linuron is designed to have greater contact action and thus control both established and germinating weeds in bulb crops. Careful **selection** of the most suitable mixture of active ingredients is therefore necessary for a particular crop/weed situation.

### Insecticides and acaricides

The insects and mites have three main points of weakness for attack by pesticides which are as follows. Their waxy exoskeletons (Figure 10.2) may be penetrated by wax-dissolving **contact** chemicals; their abdominal spiracles allow **fumigant** chemicals to enter tracheae; and their digestive systems, in coping with the large food quantities required for growth, may take in **stomach** poisons. Five main groups of insecticides are described. Details of undesirable toxicities and residues to spraying operators and the general public respectively are discussed later in this chapter.

**Dormant-season** control of pests may be achieved on trees such as apple by the use of toxic contact insecticides, e.g. tar oil. It kills eggs (**ovicide**) of aphids, red spider mite and capsids, but useful predators are also eliminated (*see* integrated control). Most other insecticides have activity on a more limited number of pests.

**Malathion** belongs to the organophosphorus group. It enters through the cuticle and, on reaching the nervous system, interferes with 'messages' crossing the nerve endings of pests such as aphids, mealy bug, flies and red spider mites. The related chemical **dimethoate** is slightly soluble in water, and may **systemically** move through the xylem and phloem tissues of the plant before being sucked up by the aphid stylet. The long lasting (**residual**) property of another ingredient, **chlorfenvinphos**, enables its use as a soil insecticide against emerging root fly larvae.

**Pirimicarb** belongs to the third (carbamate) group, is systemic, and controls many aphid species without affecting beneficial ladybirds. **Carbaryl** has stomach and contact action against caterpillars, capsids and earthworms in turf. **Aldicarb** combines a soil action against nematodes with a systemic, broad spectrum activity against foliar pests, e.g. aphids, whitefly, leaf miners, mites and nematodes of ornamental plants. This group of chemicals acts on the insect **nervous** system in a similar manner to the organophosphorus group.

The fourth, organochlorine, group of chemicals has been implicated in environmental hazards, and is progressively less used in large-scale horticulture. However, **HCH** may be used to penetrate the thick waxy cuticle of pests, e.g. vine weevil, and **dicofol** is used against several mite species.

The fifth group contains chemicals derived from plants. **Nicotine**, a natural extract from tobacco, is used as a spray or smoke, and enters the spiracles of aphids, capsids, leaf miner adults and thrips to reach their nervous system. It persists only briefly on the plant. **Pyrethrum**, an extract from a chrysanthemum sp., has been largely superseded by a range of related synthetic products, e.g. resmethrin and permethrin, which act against whitefly and caterpillars respectively. Their action is both **fumigant**, stunning flying insects, and **contact**. **Derris**, an extract from a tropical legume (and originally a fish poison), interferes with the **respiration** of a wide range of insects and mites. The insecticide **diflubenzuron** acts selectively on caterpillars and sciarid fly larvae by disturbing the insect's moulting.

### Nematicides

No active ingredients are, at present, available exclusively for nematode control. Soil-inhabiting stages of cyst nematodes, stem and bulb eelworm, and some ectoparasitic root eelworms are effectively reduced by soil incorporation of granular pesticides, e.g. **aldicarb** (above) at planting time of crops such as potatoes and onions. This group of chemicals acts systemically on leaf nematodes of plants such as chrysanthemum and dahlia.

### Fungicides

Fungicides must act against the disease but not seriously interfere with plant activity. **Protectant** chemicals prevent the entry of hyphae into roots, and the germination of spores into leaves and other aerial organs (Figure 11.2). **Systemic** chemicals enter roots, stems, and leaves, and are translocated to sites where they may affect hyphal growth and prevent spore production. Although there are many fungicidal chemical groups, three are chosen here as examples.

**Inorganic chemicals** (i.e. containing no carbon) have long been used to protect horticultural crops. **Copper** salts, when mixed with slaked lime (**Bordeaux mixture**) form a barrier on the leaf to fungi, e.g. potato blight. Fine-grained (colloidal) **sulphur** controls powdery mildews, and may be heated gently in greenhouse 'sulphur lamps' to control this disease by vapour action on plants such as roses.

**Organic chemicals** contain carbon. **Zineb** (dithiocarbamate group) and related synthetic compounds act protectively on a wide range of quite different foliar diseases, e.g. downy mildews, celery leaf spot and rusts, by preventing spore germination.

**Benomyl** (benzimidazole group) is an example of a **systemic** ingredient which moves upwards through the plant's xylem tissues, slowing hyphal growth and spore production of fungal wilts, powdery mildews and many leaf spot organisms. Damping off, potato blight and downy mildews are unaffected by this chemical group. Many different systemic groups are now used in horticulture.

About 200 active ingredients for weed, pest and disease control are available. Some are withdrawn and some introduced every year. Ministry of Agriculture and commercial literature is constantly updated to keep growers informed.

### Resistance to pesticides

The development of resistant individuals from the millions of susceptible weeds, pests and diseases occurs most rapidly when exposure to a particular chemical is continuous, or when a pesticide acts against only one body process of the organism. Resistance, e.g. in aphids, to one member, e.g. malathion, of a chemical group confers resistance to other chemicals in the same (e.g. organophosphorus) group. Horticulturists should therefore follow the strategy of alternating between different **groups** and not simply changing active ingredients. Particular care should be taken with **systemic** chemicals which present to the organism inside the plant a relatively weak concentration against which the resistant organism can develop. Increase in **dosage** of the chemical will not, in general, provide a better control against resistant strains. Biological control, unlike chemical control, does not create resistant pests.

### Formulations

Active ingredients are mixed with other ingredients to increase the efficiency and ease of application, prolong the period of effectiveness, or reduce the damaging effects on plants and man. The whole product (**formulation**) in its bottle or packet is given a **trade name**, which often differs from the name of the active ingredient.

**Liquids** (emulsifiable concentrates) contain a light oil or paraffin base in which the active ingredient is dissoved. Detergent-like substances (**emulsifiers**) present in the concentrate enable a stable emulsion to be produced when the formulation is diluted with water. In this way, the correct concentration is achieved throughout the spraying operation. Long chain molecular compounds (**wetter/spreaders**) in the formulation help to stick the active ingredients on to the leaf after spraying, particularly on smooth, waxy leaves such as cabbage. The water repellant waxy leaves of peas when sprayed with herbicides, e.g. aziprotryne, enable them to survive intact, while weeds, e.g. fat hen, are controlled.

**Wettable powders** containing extremely small particles of active ingredient and wetting agents form a stable suspension for only a short period of time when diluted in the spray tank. Continuous stirring or shaking of the diluted formulation is thus required. An inert **filler** of clay-like material is usually present in the formulation to ease the original grinding of particles, and also to help increase the shelf life of the product. It is suggested that this formulation is mixed into a thin **paste** before pouring through the filter of the sprayer. This prevents the formation of lumps which may block nozzles.

**Dusts** are applied dry to leaves or soil, and thus require less precision in grinding of the constituent particles, and less wetting agent.

**Seed dressings** protect the seed and seedling against pests and diseases. A low percentage of active ingredient, e.g. captan, applied in an inert clay-like filler or liquid reduces the risk of chemical damage to the delicate germinating seed.

**Baits** contain attractant ingredients, e.g. bran and sugar mixed with the active ingredient, e.g. methiocarb, both of which are eaten by the pest, e.g. slugs.

**Granules** formulated to a size of about 1.0 mm contain an inert filler, e.g. pumice or charcoal, on to which is coated the active ingredient. Granules may act as soil sterilants (e.g. dazomet), residual soil herbicide (e.g. dichlobenil), residual insecticide (e.g. chlorfenvinphos), or broad spectrum soil nematicide and insecticide (e.g. aldicarb). Granular formulations present less hazards to the operator and fewer spray-drift problems.

**Smokes** containing sodium chlorate, a sugar and heat resistant active ingredient, e.g. HCH, produce, on ignition, a vapour which reaches all parts of a greenhouse to fumigate against pests, e.g. whitefly.

**Labels on commercial formulations** give details of the active ingredient contained in the product. Application rates for different crops are included. The Ministry of Agriculture, Fisheries and Food (MAFF) approves pesticide products for effectiveness.

### Application of herbicides and pesticides

This subject is described in detail in machinery texts. However, certain basic principles related to the covering of the leaf and soil by sprays will be mentioned. The application of liquids and wettable powders by means of **sprayers** may be adjusted in terms of pressure and nozzle type to provide the required spray rate. **Cone** nozzles produce a turbulent spray pattern suitable for fungicide and insecticide use, while **fan** nozzles produce a flat spray pattern for herbicide application. In periods of active plant growth, fortnightly sprays may be necessary to control pests and diseases on newly expanding foliage. **High volume** sprayers apply the diluted chemical through **cone** nozzles at rates of 600–1000 l/ha in order to cover the whole leaf surface with droplets of 0.04–0.10 mm diameter. Cover of the underleaf surface with pesticides may be poor if nozzles are not directed horizontally or upwards. Soil applied chemicals, e.g. herbicides or drenches, may be sprayed at a larger droplet size, 0.25–0.5 mm in diameter, through a selected **fan** nozzle. Correct height of the sprayer boom above the plant essential for downward directed nozzles if the spray pattern is to be evenly distributed. Savings can be achieved by **band spraying** herbicides in narrow strips over the crop to leave the inter-row for mechanical cultivation.

**Medium volume** (200–600 l/ha) and **low volume** (50–200 l/ha) equipment, such as knapsack sprayers, apply herbicides and pesticides on to the leaf at a lower droplet density, and in tree crops, mist blower equipment creates turbulence, and therefore increased spray travel, by means of a power driven fan. **Electrically charged** nozzles, which transfer a strong charge to the sprayed droplet, are being developed. The droplet is attracted to the target plant and very little spray lands on the soil. **Ultra low volume** sprays (up to

50 l/ha) are dispersed on leaving the sprayer by a rapidly rotating disc which then throws regular sized droplets into the air. Larger droplets (about 0.2 mm) are created by herbicide sprayers to prevent spray-drift problems, while smaller droplets (about 0.1 mm) allow good penetration and leaf cover for insecticide and fungicide use.

**Fogging machines** used in greenhouses and stores produce very fine droplets (about 0.015 mm diameter) by thermal and mechanical methods, and use small volumes of concentrated formulation (less than 1 litre in 400 cubic metres) which act as fumigants in the air, and as contact pesticides when deposited on the leaf surface. **Dust** and **granule applicators** spread the formulations evenly over the foliage, or ground surface. When mounted on seed drills and/or fertilizer applicators, granules may be incorporated into the soil. Care must be taken to ensure good distribution to prevent pesticide damage to germinating seeds or planting material.

### Integrated control

Integrated control requires the grower to understand all types of control measure, particularly biological and chemical, in order that they complement each other. In greenhouse production of cucumbers, the *Encarsia formosa* parasite and *Phytoseiulus persimilis* predator are used for whitefly and red spider mite control respectively. However, the other harmful pest and disease species must be controlled without killing the parasites and predators. Drenches of systemic insecticide, e.g. pirimicarb against aphids, soil insecticides, e.g. HCH against thrips pupae, and systemic fungicide drenches, e.g. benomyl against wilt diseases and powdery mildew, are all applied away from the sensitive biological control organisms. Similarly, high-volume sprays of selective chemicals, e.g. dichlofluanid against grey mould, *Bacillus thuringiensis* extracts against caterpillars, have little or no effect on the parasite and predator. Similar considerations may be given in control of apple pests and diseases. Reduced

usage of extremely toxic winter washes, and **selective** caterpillar and powdery mildew control by chemicals, e.g. diflubenzuron and fenarimol respectively, allow the unhindered build-up of beneficial organisms, e.g. predatory capsids and mites.

The organic methods of horticulture emphasize the non-chemical practices in plant protection (as well as in soil fertility). Hedges are developed within 100 metres of production areas and are clipped only one year in four to maintain natural predators and parasites. Rotations are closely followed to enable soil-borne pest or disease decline, while encouraging soil fertility. Resistant cultivars of plant are chosen, and judicious use of mechanical cultivations and flame weeding enables pests, diseases and weeds to be exposed or buried. A restricted choice of pesticide products such as pyrethrins, derris, metaldehyde (with repellant), sulphur, copper salts, and soft soap are allowed to be applied should the need arise. *Bacillus thuringiensis* extract, and also pheromone attractants, are similarly used. Table 12.1 gives a list of permitted substances.

### Supervised control

Most plants can tolerate low levels of pest and disease damage without yield reduction. Thus, cucumbers require more than 30% leaf area affected by red spider mite before economic damage occurs in terms of yield loss. Damage assessments are used in apple orchards to decide whether control measures are necessary. Thus, at green cluster stage (before flowers emerge) chemical sprays are considered only when an average of half the observed buds have five aphids/bud. Similarly, three winter moth larvae per bud-cluster merit control at late blossom time. **Pheremone traps** enable the precise time of maximum codling moth emergence to be determined in early June. Catches of less than ten moths per trap per week do not warrant control. The Ministry of Agriculture issued **spray warning** information to growers when serious pests, e.g. carrot root fly,

**Table 12.1** Permitted products for plant pest and disease control in organic crop production

---

Preparations on basis of metaldehyde containing a repellent to higher animal species and as far as possible applied within traps

Preparations on basis of pyrethrins extracted from *Chrysanthemum cinerariaefolium*, containing possibly a synergist

Preparations from *Derris elliptica*

Preparations from *Quassia amara*

Preparations from *Ryania speciosa*

Propolis

Diatomacious earth

Stone meal

Sulphur

Bordeaux mixture

Burgundy mixture

Sodium silicate

Sodium bicarbonate

Potassium soap (soft soap)

Pheromone preparations

*Bacillus thuringiensis* preparations

Granulose virus preparations

Plant and animal oils

Paraffin oil

---

and diseases, e.g. potato blight, are likely to occur. Supervised control may greatly reduce pesticide costs.

**Safety with herbicides and pesticides**

The basic biochemical similarities between all groups of plants and animals demand careful examination of active ingredients before their commercial release; otherwise serious damage or death to non-target organisms may result.

Regulations under the Food and Environmental Protection Act 1985 (**FEPA**) control the approval of pesticide products in terms of their safety to humans and wildlife. The lethal dose figure ($LD_{50}$), expressing the amount of active ingredient in milligrams which kills one kilogram of animal tissue, dictates to a large extent the precautions needed for a grower to safely mix and apply a product. The lower the $LD_{50}$ figure for a chemical, the more toxic it is. The label present on each packet or bottle of pesticide must provide detailed instructions with regard to precautions and protective clothing required for a chemical's safe usage. The amount of protection stated on the label commonly reflects the $LD_{50}$ status of the toxic ingredient. Highly toxic chemicals, e.g. the soil sterilant methyl bromide ($LD_{50} = 1$) should be applied only by specially trained personnel; a less toxic ingredient, e.g. the fungicide captan ($LD_{50} = 9000$), requires a coverall, face shield and protective gloves when dealing with the concentrated product.

Although acute toxicity is the main consideration in the safe use of pesticides, other properties must be borne in mind. The residual nature of certain chemicals, e.g. the acaricide endosulfan, on edible crops, necessitates a **spray-to-harvest interval** of at least six weeks. A shorter interval is allowed for some other pesticides; e.g. one week for the insecticide dimethoate, and one day for the insecticide dichlorvos. Certain active ingredients with low residual properties, e.g. zineb, nonetheless have a long (three week) interval when used on large-surface-area crops such as lettuce. Taint problems in certain chemical/crop combinations such as HCH with potatoes, require an application-to-planting interval of at least eighteen months.

The maximum residue levels (**MRL**) of pesticide in foodstuffs is strictly controlled. In lettuce, for example, the level of the insecticide HCH must not exceed two parts per million.

The Control of Substances Hazardous to Health Regulations 1988 (**COSHH**) require the horticulturist to assess whether each pesticide application is necessary. They also require the use of the safest ingredients, equipment and protective clothing to reduce the risk to humans.

**Health and Safety Regulations 1975**, summarized in the government 'Poisonous Chemicals on the Farm' leaflet, specify the correct procedures for pesticide use. A detailed register must be kept of spraying operations and any dizziness or illness

reported. Correct washing facilities must be provided. A lockable dry store is necessary to keep
chemicals safe. Warnings of spraying operations
should be prominently displayed. Suitable first
aid measures should be available. Used pesticide
containers should be burnt or buried in a safe
place.

**Protective Clothing** appropriate to the pesticide
should be worn. A one-piece fabric coverall suit
with a hood protects most of the body from diluted
pesticide. Rubberized suits may be used in conditions of greater danger, e.g. in an enclosed
greenhouse environment when dealing with ultra-
low-volume spray, or when applying upward-
directed sprays into orchards. Rubber boots
should be worn inside the legs of the suit. Thick
gauge gauntlets are worn ouside the suit when
dealing with concentrates, but inside when
spraying. Face shields should be worn when mixing toxic concentrates. A face mask covering the
mouth and/or nose is capable of filtering out less
toxic active ingredients from sprays, but a **respir-
ator** with its large filter is required for toxic products, particularly when used in greenhouses where
toxic fume levels build up.

**Wildlife**. Pesticides should not be sprayed near
ponds and streams unless designed for aquatic
weed control. Crops frequented by **bees**, e.g.
apples and beans, should be sprayed with insecticides only in the evening when most of the insects'
foraging has ceased. Bee-keepers should be
informed of spraying operations. Very long-lasting
insecticides, e.g. HCH, should not be used where
suitable alternatives are available, particularly in
large scale plant production, since small insects
and larger animals, e.g. hawks, in the **food webs**
accumulate the chemicals to toxic levels in their
bodies and eggs. Public and government awareness of pesticide dangers have resulted in more
wide-ranging legislation under Section III of the
Food and Environmental Protection Act 1985.
This requires that chemical manufacturers, dis-
tributors and professional horticulturists show
demonstrable skills in the choice and application
of pesticides. The Act also seeks to make information about pesticides available to the public.

**Phytoxicity**

Phytotoxicity (or plant damage) may occur when
pesticides are unthinkingly applied to plants. Soil-
applied insecticides, e.g. aldicarb, cause pot
plants, e.g. begonias, to go yellow if used at more
than the recommended rate. Foliage-applied
fungicides, e.g. dinocap, cause some chrysanthemum cultivars to produce unsightly yellow
spots on their foliage. Careful examination of the
pesticide (particularly **herbicide**) packet labels
often prevents this form of damage.

## PRACTICAL EXERCISES

1    Compare the relative levels of success
when annual weeds are hoed in dry and wet
conditions.
2    Observe the low incidence of pests and diseases
on a range of wild plants.
3    Follow the progress of a biological control
programme against greenhouse pests. Assess
the suitability of pesticides used for an integrated control programme.
4    Compare the effectiveness of a range of soil-
applied herbicides in spring, using an area
where a wide range of weeds had previously
occurred. Record which weed species survive
each treatment.
5    Assess the proportion of pests killed after a
systemic pesticide is drenched on the soil.
6    Use a sprayer filled with dilute dye solution to
illustrate the spray pattern produced on white
card.

## FURTHER READING

British Crop Protection Council. *The UK Pesticide
Guide* (1993).
Fryer, J.D., and Makepeace, R.J. *Weed Control
Handbook*, Vol. 2 (Blackwell Scientific Publications,
1978).
Hance, R.J., and Holly, K. *Weed Control Handbook*,
Vol. 1 (Blackwell Scientific Publications, 1990).

Hussey, N.W., and Scopes, N. *Biological Pest Control* (Blandford Press, 1985).

MAFF. Pesticides 1990: *Pesticides Approved under the Control of Pesticides Regulations 1986* (1986).

Matthews, G.A. *Pesticide Application Methods* (Longman, 1979).

Scopes, N., and Stables, L. *Pest and Disease Control Handbook* (British Crop Protection Council, 1989).

Savigear, E. *Garden Pests and Predators* (Blandford, 1992).

Temple, J. *Gardening Without Chemicals* (Thorson, 1986).

*Welcome to the World of Environmental Products* (Rhone-Poulenc, 1992).

*Working with Pesticides – A Guide* (Schering Chemicals, 1987).

# 13

# Soil as a Growing Medium

*The plant takes up water and nutrients from the growing medium through its roots which also provide anchorage (see root structure). In order to secure water throughout the season the roots penetrate deep into the soil. The plant nutrients are extracted from a very dilute soil solution and so the roots must explore extensively in the top layers to maintain nutrient supplies. Consequently the roots are normally very extensive and with a shape that brings the maximum absorptive surface into contact with soil particle surfaces around which is the water and nutrients. Within one growing season a single plant growing in open ground develops some 700 km of root which has a surface area of 250 m$^2$; 700 m$^2$ if the surface area includes the root hairs. The vast root system that develops is usually more than is required to supply the plant in times of plenty but the extent of the root system is indicative of what the plant needs to protect itself against unfavourable conditions. It also serves to remind us of what we undertake to provide when we restrict growth accidently or deliberately. When growing in containers there is the opportunity to provide an ideal growing medium but the continued supply of water and nutrients becomes more critical (see Chapter 17).*

*In this chapter the nature of soil is detailed. Its qualities as a root environment are explored and the basis of managing it as a growing medium for horticultural plants is established.*

## ROOT REQUIREMENTS

The growing tip of the root wriggles through the growing medium following the line of least resistance. Roots are able to enter cracks which are, or can be readily opened up to about 0.2 mm in diameter. Compacted soils severely restrict **root exploration** but once into these narrow channels the root is able to overcome great resistance to increase its diameter. Anything which reduces root exploration and activity can limit plant growth. When this happens action must be taken to remove the obstruction to root growth or to supply adequate air, water and nutrients through the restricted root volume.

In order to grow and take up water and nutrients the root must have an energy supply. A constant supply of energy is only possible so long as oxygen is brought to the site of uptake (*see* respiration). Consequently the soil spaces around the root must contain air as well as water. A lack of oxygen or a build-up of carbon dioxide will reduce the root's activity. Furthermore, in these conditions anaerobic bacteria will proliferate and many produce toxins such as ethylene. In warm summer conditions roots can be killed back after one or two days in waterlogged soils.

## COMPOSITION OF SOILS

Mineral soils form in layers of rock fragments over the Earth's surface. They are made up of mineral matter comprising sand, silt and clay particles and organic matter which is the part derived from living organisms. This framework of solid material retains water and gases in the gaps or **pore space**. The water contains dissolved materials including plant nutrients and oxygen and is known as the **soil solution**. The **soil atmosphere** normally comprises nitrogen, rather less oxygen and rather more carbon dioxide than in normal air and traces of other gases. Finally, a soil capable of sustaining plants is alive with microorganisms.

The composition of a typical mineral soil is given in Figure 13.1 which also illustrates the variation that can occur. The content of the pore space varies continually as the soil dries out and is rewetted. The spaces can be altered by the compaction or 'opening up' of the soil which in turn has a significant effect on the proportions of air and water being held. Over a longer period the organic matter level can vary. The composition of the soil can be influenced by many factors and under cultivation these have to be managed to provide a suitable root environment.

Organic soils have a considerably higher organic matter content and are dealt with in Chapter 15.

## SOIL FORMATION

The Earth formed from a ball of molten rock minerals. The least dense rocks floated on the top and as they cooled the surface layer of granite, with basalt just below, solidified to form the Earth's crust. The Earth's surface has had a long and turbulent history during which it has frequently fractured, crumpled, lifted and fallen, with more molten material being pushed up from

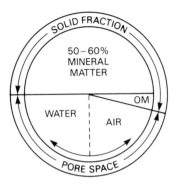

**Figure 13.1** *Composition of a typical cultivated soil. The solid fraction of the soil is made up of mineral (50−60%) and organic (1−5%) matter. This leaves a total pore space of about 40% which is filled by air and water, the proportions of which vary constantly.*

below through the breaks in the crust and in volcanoes.

### Weathering and erosion

**From the moment rocks are formed and exposed to the elements they are subjected to weathering, the breakdown of rocks, and erosion, the movement of rock fragments and soil.**

Chemical weathering is mainly brought about by the action of carbonic acid which is produced wherever carbon dioxide and water mix, as in rainfall. Some rock minerals dissolve and are washed away. Others are altered by various chemical reactions, most of which occur when the rock surface is exposed to the atmosphere. All but the inert parts of rock are eventually decomposed and the rock crumbles as new minerals are formed and soluble material is released. Oxidation is particularly important in the formation of iron oxides which give soils their red and yellow, or blue and grey colours.

Physical or mechanical weathering processes break the rock into smaller and smaller particles without any change in the chemical character of the minerals. This occurs on exposed rock surfaces along with chemical weathering but, in contrast, has little effect on rocks protected by layers of soil. The main agents of physical weathering are frost, heat, water, wind and ice. In temperate zones, frost is a major weathering agent. Water percolates into cracks in the rock and expands on freezing. The pressures created shatter the rock and as the water melts a new surface is exposed to weathering. In hot climates the rock surface can become very much hotter than the underlying layers. The strains created by the different amounts of expansion and the alternate expansion and contraction cause fragments of rock to flake off the surface; this is sometimes known as the 'onion skin' effect. Moving water or wind carries fragments of rock which rub against other rocks and rock fragments, wearing them down. Where there are glaciers the rock is worn away by the 'scrubbing brush' effect of a huge mass of ice loaded with stones and boulders bearing down on the underlying rock.

Biological weathering is attributable to organisms which fragment rock by both chemical and physical means, e.g. they produce carbon dioxide which in conjunction with water forms carbonic acid; roots penetrate cracks in the rock and as they grow thicker they exert pressure which further opens up the cracks.

### Igneous rocks

Igneous rocks are those formed from the molten material of the Earth's crust. All other rock types, as well as soil, are ultimately derived from them. When examined closely most igneous rocks can be seen to be a mixture of crystals. **Granite**, for example, contains crystals of quartz, white and shiny; felspars which are grey or pink; and micas which are shiny black.

As granite is weathered, 'rotted', the felspars are converted to kaolinite (one of the many forms of clay) and soluble potassium, sodium and calcium which are the basis of plant nutrition. Similarly, the mica present is chemically changed. Whilst much of the potassium is retained by the clays, the other soluble material is carried by water to the sea making the sea 'salty'. The inert quartz grains are released and form sand grains.

### Sedimentary rock

Sedimentary rock is derived from accumulated fragments of rock. Most have been formed in the sea or lakes to which weathered rock is carried by agents of erosion. Layers of sediment build up and the great depths of sediment eventually become rock strata. In subsequent earth movements much of it has been raised up above sea level and weathered again.

Moving water and winds are able to carry rock particles and are thus important agents of erosion. As their velocity increases, the '**load**' they are able to carry increases substantially. The fast-moving water in streams is able to carry large particles but in the slower-moving rivers some of the load is dropped. The particles settle out in order of size (*see* settling velocities). This leads to the **sorting** of rock fragments, i.e. material is moved and deposited according to particle size. By the time the rivers have reached the sea or lakes only the finest sands, silts, and clays are in the water. As the river slows on meeting the sea or lake all but clay is dropped. The clay eventually settles slowly in the still water of the sea or lake. Moving ice is also an agent of erosion but the load dropped on melting consists of unsorted particles known as **boulder clay** or till.

The type of sedimentary rock formed depends on the nature of its ingredients. Sandstones, siltstones, and mudstones are examples of sedimentary rocks derived from sorted particles in which characteristic layering is readily seen. Limestones are formed from the accumulation of shells or the precipitation of materials from solution mixed with varying amounts of deposited mud. Chalk is a particularly pure form derived from the calcium carbonate remains of minute organisms.

## Metamorphic rock

Metamorphic rock is formed from igneous or sedimentary rocks which have been altered by the extreme pressures and temperatures associated with movements and fracturing in the Earth's crust or the effect of huge depths of rock on underlying strata over very long periods of time. Slate is formed from shales, quartzite from sandstone, and marble from limestone. Metamorphic rock tends to be more resistant to weathering than the original rock.

## SOIL DEVELOPMENT

**Soil development occurs in the loose rock fragments overlying the Earth's crust. This is the parent material which has an important effect on the nature of the soil formed. However, the soil formed is influenced by climate, topography, vegetation, animals including man, and time.**

**Topography** is the detailed description of land features which are the result of the interaction between the underlying rocks, the agents of weathering and erosion, and time. The landscape may look as though it is quite stable but even dramatic mountain ranges are worn down to flat plains: a process that has occurred several times in the history of the Earth following the numerous crustal disturbances.

On rocks and rock debris plant growth begins with mosses and lichens which help to stabilize the loose material by binding the surface and reducing wind speed over the surface. The dead plants become incorporated in the young soils and their breakdown products also help hold it together. Nutrients are held in the top surface and recycled by a complex colony of soil organisms instead of being washed away (*see* nutrient cycles). As the young soil deepens, larger plants appear. Their roots help stabilize the soil further and, along with the carbonic acid they produce, extend the depth of weathering. The process of soil formation, which takes thousands of years, speeds up as the particles become more finely divided and the living organisms become established.

## Sedentary soils

Sedentary soils develop in the material gradually weathered from the underlying rock. True sedentary soils are uncommon because most loose rock is eroded, but the same process can be seen where great depths of transported material have formed the parent material, as in the boulder clays left behind after the Ice Ages. A hole dug in such a soil shows the gradual transition from unweathered rock to an organic matter rich topsoil (Figure 13.2). Under cultivation a distinct topsoil develops in the plough zone.

## Transported soils

Transported soils are those that form in eroded material that has been carried from sites of weathering sometimes many hundreds of miles away from where deposition has occurred e.g. by glaciers. They can be recognized by the definite boundary between the eroded material and the underlying rock and associated rock fragments. Where more than one soil material has been transported to the site, as in many river valleys, several distinct layers can be seen. The right-hand part of Figure 13.2 shows an example.

Raindrops striking soil dislodge loose particles which tend to move downhill. As a result, surface soil is slowly removed from higher ground and accumulates at the bottom of slopes. This means that soils on slopes tend to be shallow, whereas at the bottom of the slopes deep, transported soils develop, known as colluvial soils. Material washed away in running water eventually settles out according to particle size. The river valley bottoms become covered with material, called alluvium, in which **alluvial** soils develop. Wind removes dry sands and silts which are not 'bound in' to the soil and large areas have become covered with

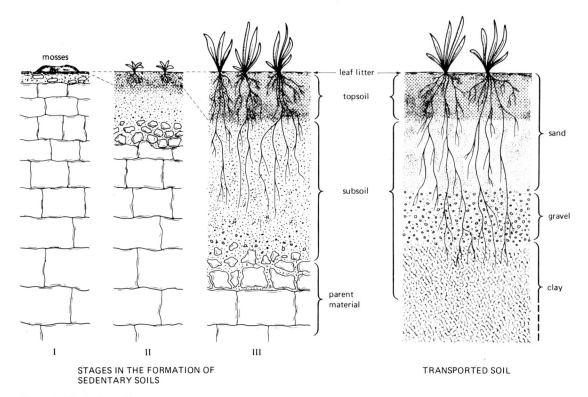

**Figure 13.2** *Soil profiles*
*The development from a young soil consisting of a few fragments of rock particles to a deep sedentary soil is shown alongside a transported soil. A subsoil, topsoil and leaf litter layer can be identified in each soil.*

wind-blown deposits known as 'loess' or 'brick-earth'. Many of these transported soils provide ideal rooting conditions for horticultural crops because they tend to be deep, loose and open. Most are easily cultivated. Those that have a high silt or fine sand content, notably the brickearths, may be prone to compaction.

## SOIL COMPONENTS

The solid parts of soils consist of mineral matter derived from rocks and organic matter derived from living organisms. Levels of organic matter are dealt with in Chapter 15. **Most cultivated soils have predominantly mineral particles which vary enormously in size from boulders, stones, and gravels down to sand, silt, and clay.**

**Particle size classes**

There is a continuous range of particle sizes but it is convenient to divide them into classes. Three major classification systems in use today are those of the International Society of Soil Science (ISSS), United States Department of Agriculture (USDA) and the Soil Survey of England and Wales (SSEW). *See* Figures 13.3 and 13.4. In the text the SSEW scale used by the Agricultural Development and Advisory Service of England and Wales (ADAS) is adopted. In each case, soil is considered to consist of those particles that are less than 2 mm in diameter. The silt and clay particles are sometimes referred to as 'fines'.

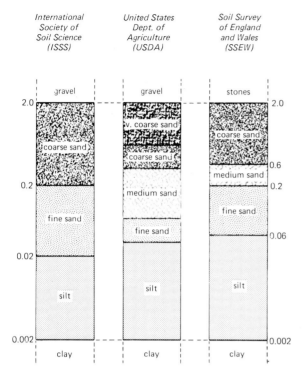

**Figure 13.3** *Particle size classes.*

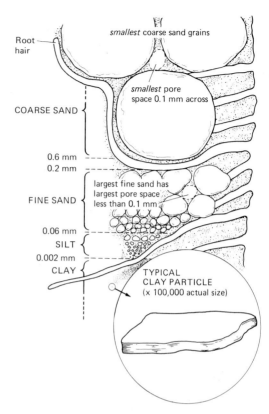

**Figure 13.4** *Soil particle sizes*
*The relatives sizes of coarse sand, fine sand, silt and clay particles (based on SSEW classification) with root hairs drawn alongside for comparison. Note that even the smallest pore spaces between unaggregated spherical coarse sand grains still allow water to be drawn out by gravity and allow some air in at field capacity, whereas most pores between unaggregated fine sand grains remain water-filled (pores less than 0.1 mm diameter).*

### Sand

**Sand grains are particles between 0.06 mm and 2.0 mm in diameter. They are gritty to the touch; even fine sand has an abrasive feel. Sand is mainly composed of quartz.** The shape of the particles varies from the rough and angular sand to more weathered, rounded grains. They are frequently coated with iron oxides, giving sand colours from very pale yellow to rich, rusty brown. Silver sand has no iron oxide covering. Chemically most of the sand grains are **inert**; they neither release nor hold on to plant nutrients and they are not cohesive.

The influence of sand on the soil is mainly physical and as such the size of the particles is the important factor. As the particles become smaller and the volume of individual grains decreases, the **surface area** of the same quantity of sand becomes greater. This is readily demonstrated by taking a cube and then cutting it up into smaller cubes

(Figure 13.5). While the total volume is the same in all the small cubes compared with the original large cube, the sides of each are smaller but the total surface area is much greater because new surfaces have been exposed. Many soil particle characteristics are directly related to particle size and in particular to surface area. Sand grains are non-porous so their water-holding properties are directly related to their surface areas. It can be readily seen that, since water will not flow through gaps less than about 0.1 mm in diameter, there

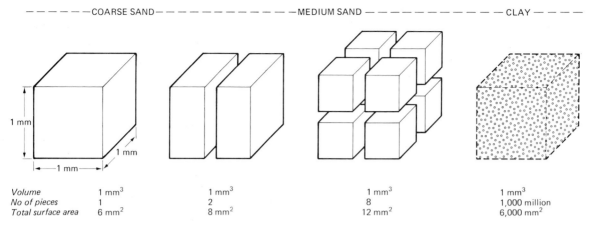

**Figure 13.5** *Surface area of soil particles*
*The effect of sub-dividing a cube corresponding in size to a grain of coarse sand. The same volume of medium sand is made up of over eight times more pieces which have a total surface area more than double that of coarse sand. It requires over a thousand million of the largest clay particles to make up the volume of one grain of coarse sand and their surface area is approximately one thousand times greater.*

are very big differences in the drainage characteristics of coarse and fine sands (Figure 13.4). Consequently, soils dominated by coarse sand are usually free draining but have poor water retention whereas those composed mainly of unaggregated fine sand hold large quantities of water against gravity. The water held on all sand particles is readily removed by roots (*see* available water).

**Clay**

**Clay particles are those less than 0.002 mm in diameter. There are many different clay minerals, e.g. kaolinite, montmorillinite, vermiculite, and mica, all of which are derived from rock minerals by chemical weathering.** Most clay minerals have a layered crystalline structure and are platelike in appearance (Figure 13.4). The clay particles have surface charges which give clay its very important and characteristic property of **cation exchange**. The charges are predominantly negative which means that the clay platelets attract positively charged **cations** in the soil solution. These include the nutrients potassium, as $K^+$; ammonia, as $NH_4^+$; magnesium, as $Mg^{++}$; and calcium, as

$Ca^{++}$, as well as hydrogen and aluminium ions. These ions are held in an **exchangeable** way so that they remain available to plants but are prevented from being leached unless displaced by other cations. The greater the **cation exchange capacity**, the greater the reserves of cations held this way.

Hydrogen and aluminium cations make the soil acid. The other cations $Ca^{++}$, $Mg^{++}$, $K^+$ and $Na^+$ are called **bases** and make soils more alkaline (*see* soil pH). The proportion of the cation exchange capacity occupied by bases is known as its **percentage base saturation**. A soil's **buffering capacity**, i.e. its ability to resist changes in soil pH, also depends on these surface reactions. The presence of high levels of exchangeable aluminium and hydrogen means that very large quantities of calcium, in the form of lime, are required to raise the pH of acid clays. In contrast only small quantities of lime are needed to raise the pH of a sand by the same amount (*see* liming).

The clay particles are so small that the minute electrical forces on the surface become dominant (Figure 13.5). Thus clay and water mixtures behave as colloids. This gives clay soils properties of cohesion, plasticity, shrinkage and swelling.

The small particles can pack and stick *together* very closely and in a continuous mass they restrict water movement. The water-holding capacity of clay-dominated soils is very high because of the large surface areas and because many of the particles are porous. However, a high proportion of the water is held too tightly for roots to extract (*see* available water). Moist balls of clay are plastic, i.e. can be moulded. On drying, they harden and some shrink. In the soil the cracks which form on shrinking become an important network of drainage channels. The cracks remain open until the soils are re-wetted and the clay swells. Humus and calcium appear to combine with clay in such a way that when the combination dries, extensive cracking occurs and favourable growing conditions result. Some clays are non-shrinking and are consequently more difficult to manage.

### Silt

**Silt particles are those between 0.002 mm and 0.06 mm in diameter. Most of this size range are inert and non-porous like sands but many particles, including felspar fragments, have the properties of clay.** Soils dominated by silt do have a small cation exchange capacity but in the main they behave more like a very fine sand. They have very good water-holding capacity and a high proportion of this water can be taken up by plants. When wetted they have a distinctly soapy or **silky** feel. Silt soils are made up of particles that readily pack closely together but have little ability to form stable crumbs (*see* aggregates). This makes them particularly difficult to manage.

### Stones and gravel

Particles larger than 2 mm in diameter are known variously as grit, gravel, stones and boulders, according to size. The effect of stones on cultivated areas depends on the type of stone, size, and the proportion in the soil. The presence of even a few stones larger than 20 cm prevents cultivation.

Stones in general have detrimental effects on mechanized work; ploughshares, tines, and tyres are worn more quickly, especially if the stones are hard and sharp such as broken flint. Stones interfere with drilling of seeds and the harvesting of roots. Close cutting of grasses is more hazardous where there are protruding stones. Mole draining becomes less effective in stony soils and large stones make it impossible. Stones can accumulate in layers and become interlocked to form stone pans. In very gravelly soils the water-holding capacity is much reduced and the increased leaching leads to acid patches. Nutrient reserves are also reduced by the dilution of the soil with inert material. Stones can help water infiltration, protect the surface from capping, and check erosion by wind or water.

### SOIL TEXTURE

Soil texture describes the composition of a soil. In most cultivated soils the mineral content forms the framework and exerts a major influence on its characteristics.

**Although there is no universally accepted definition of soil texture, it can usefully be defined as the relative proportions of the sand, silt and clay particles in the soil.** Examples of different textures are given in Figure 13.6. Texture provides a useful guide to a soil's potential. Fine-textured soils such as clays, clay loam, silts and fine sands have good water-holding properties, whereas coarse-textured soils have low water-holding capacity but good drainage. This also means that **soil temperatures** are closely related to soil texture because water has a very much higher specific heat value than soil minerals. Consequently freely-draining, coarse sand warms up more quickly in the spring but is also more vulnerable to frosts than wetter soils.

Soils with high clay contents have good general nutrient retention, whereas nutrients are readily lost from sandy soils, especially those with a high coarse sand fraction. The application rate of pesticides and herbicides is often related to soil

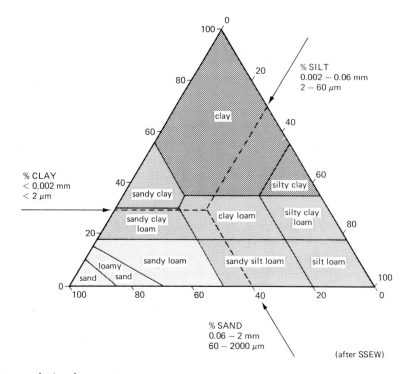

**Figure 13.6** *Soil textural triangle*
*The soil texture can be identified on this type of chart when at least two of the major size of fractions are known, e.g. 40% sand, 30% silt and 30% clay is a clay loam (SSEW Soil-Particle Size Classification).*

texture (*see* application rates). The power requirement to cultivate a clay soil is very much greater than that for a sandy soil. The expression 'heavy' for clay and 'light' for sandy soils is derived from this difference in working properties rather than the actual weight of the soil. The texture of a soil also influences the soil structure and soil cultivations.

In general the texture of a soil can be considered to be a fixed character. The addition of a calcareous clay to a sandy topsoil, a practice known as **marling**, can improve its water-holding capacity as well as reducing wind erosion, but it requires the incorporation of 500 tonnes of dry clay per hectare to convert it to a sandy loam. The practice of adding clay is now largely confined to the building of cricket squares. To 'lighten' a clay loam topsoil to a sandy loam more than 2000 tonnes of dry sand is needed on each hectare. The addition of smaller quantities of sand is often an expensive exercise to no effect: at worst it can make the resultant soil more difficult to manage.

**Mechanical analysis of soils**

Soils texture can be determined by finding the particle size distribution. There are several methods but all depend on the complete separation of the particles, the destruction of organic matter, and the removal of particles greater than 2 mm in diameter. The stones, coarse sand, medium sand and fine sand fractions can be separated by **sieving**.

Finer particles are usually separated by taking advantage of their different **settling velocities** when in suspension. The settling velocity of a particle depends on the viscosity and density of the liquid, the density and radius of the particles, and the acceleration due to gravity. If it is assumed that soil particles are spherical and have the same density and if investigations are conducted in water at 20°C, the soil particles will settle according to their size. Particles which are less than 0.001 mm in diameter are kept permanently in suspension by the bombardment of vibrating water molecules and are referred to as **colloids**, e.g. most clay particles. All particles greater than 0.002 mm will have fallen more than 10 cm after eight hours and so a sample taken at that depth can be used to calculate the clay content. Similarly, other fractions can be calculated until the sand, silt and clay are determined. The soil texture can be deduced from this information using a **textural triangle** (Figure 13.6), which is the basis of identifying **soil types**.

### Texturing by feel

A more practical method of determining soil texture, especially in the field, is by feel. This can, with experience, be a very accurate means of distinguishing between over thirty categories.

A ball of soil about the size of a walnut is moistened and worked between the fingers to remove particles greater than 2 mm and to break down the soil crumbs. It is essential that this preparation is thorough or the effect of the silt and clay particles will be masked. **Sands** are soils which have little cohesion. **Sand** has little tendency to bind even when wetted and it cannot be rolled out into a 'worm'. **A loamy sand** has sufficient cohesiveness to be rolled into a 'worm' but it readily falls apart. **Loams** mould readily into a cohesive ball and what used to be known as **a loam** has no dominant feel of grittiness, silkiness or stickiness. If grittiness is detected and the ball is readily deformed it is a **sandy loam**. If it is readily deformed but has a silky feel it is a **silty loam**. **Clay loams** bind together

strongly, do not readily deform, and take a polish when rubbed with the finger. **Clays** bind together and are very difficult to deform. A **clay** readily takes a very marked polish. If there is also a feeling of silkiness it is a **silty clay**; if grittiness, it is a **sandy clay**. Wherever grittiness is detected, the designation sand can be further qualified by stating whether it is coarse, medium or fine sand, e.g. coarse sandy loam. Table 13.1 shows the range of textural groupings commonly used.

Determining texture by feel has the limitation that the influence of organic matter and chalk cannot be eliminated. Chalk tends to give a soil a silky or gritty feel but the fact that a soil is known to be chalky should not influence the texturing. Its textural class may be prefixed 'calcareous', e.g. calcareous silty clay. **Organic matter** tends to increase the cohesiveness of light soils, reduce the cohesiveness of heavy soils, and large quantities can impart a silky or greasy feel. The prefix 'organic' can be used for describing mineral soils with 6–20% organic matter. Soils with 20–35% organic matter are **peaty loams**, 35–50% organic matter **loamy peats** and soils with greater than 50% organic matter are termed **peat** (*see* organic soils).

## SOIL STRUCTURE

Soil structure **is the arrangement of particles in the soil**. In order to provide a suitable root environment for cultivated plants the soil must be constructed in such a way as to allow good gaseous exchange whilst holding adequate reserves of available water. There should be a high water infiltration rate, an interconnected network of spaces allowing roots to find water and nutrients without hindrance and free drainage. There should be no large cavities which prevent thorough contact between soil and roots and allow roots to dry out in the seedbed. The soil should be managed so that erosion is minimized. Good structural stability should be maintained so that the structure does not deteriorate and limit crop growth.

**Table 13.1**    Soil texture classification based on hand-texturing

| Textural class | Symbol | Textural group* | 'Sands' 'Other'† |
|---|---|---|---|
| *Coarse sand* | CS | | |
| *Sand* | S | | |
| *Fine sand* | FS | Sands | |
| *Very fine sand* | VFS | | |
| *Loamy coarse sand* | LCS | | |
| | | | 'Sands' |
| *Loamy sand* | LS | | |
| *Loamy fine sand* | LFS | Very light soils | |
| *Coarse sandy loam* | CSL | | |
| *Loamy very fine sand* | LVFS | | |
| *Sandy loam* | SL | Light soils | |
| *Fine sandy loam* | FSL | | |
| *Very fine sandy loam* | VFSL | | |
| *Silty loam* | ZyL | Medium soils | |
| *Loam* | L | | |
| *Sandy clay loam* | SCL | | |
| *Clay loam* | CL | | 'Other' |
| *Silt loam* | ZyCL | Heavy soils | |
| *Silty clay loam* | ZyCL | | |
| *Sandy clay* | SC | | |
| *Clay* | C | Very heavy soils | |
| *Silty clay* | ZyC | | |

\* Commonly used for determining soil-acting herbicide
  application rates.
† Commonly used for determining fertilizer application or
  flooding levels.

## Porosity

The plant roots and soil organisms live in the pores between the solid components of the growing medium. In the same way that a house is mainly judged by the living accommodation created by the bricks, wood, plaster, cement etc., so a soil is evaluated by examining the spaces created. In general the smallest pores, **micropores**, contain only water which rarely dries out and is unavailable to plants (*see* permanent wilting point). The middle-sized pores, **mesopores**, contain water available to plants and the air moves in as it is removed by plant roots. Pores greater than about 0.1 mm in diameter, called **macropores**, drain easily to allow in air within hours of being fully wetted. Ideally there should be sufficient mesopores to ensure good retention of available water, but sufficient macropores to allow free drainage, gaseous exchange, and thorough root exploration as shown in Figure 13.7.

**The key to managing most growing media is in maintaining a high proportion of air-filled pores without restricting water supply. An important indicator of a satisfactory growing medium is its air-filled porosity or air capacity, i.e. the percent-**

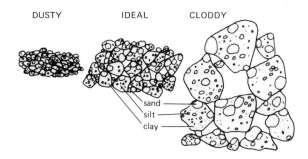

DUSTY    IDEAL    CLODDY

sand
silt
clay

**Figure 13.7** *Tilth*
*The ideal tilth for most seedbeds is made up of soil aggregates between 0.5 and 5 mm diameter. Within these crumbs are predominantly small pores (less than 0.1 mm) which hold water and between the crumbs are large pores (greater than 0.1 mm) which allow easy water movement and contain air when soil is at field capacity. (×3 actual size).*

age volume filled with air when it has completed draining.

**Bulk density** is the mass of soil per unit volume and it can be measured by taking a core of soil of known volume and weighing it after thorough drying. In normal mineral soils results are usually between 1.0 and 1.6 g/ml. The difference is largely attributable to variation in total pore space. Finer textured soils tend to have more pore space and therefore lower bulk density than sands but for all soils higher values indicate greater packing or **compaction**.

This information is not only useful to diagnose compaction problems but can also be used to calculate the weight of soil in a given volume. Assuming a cultivated soil to have a bulk density of 1.0 g/ml, the weight of dry soil in one hectare to a plough depth of 15 cm is 1500 tonnes; when compacted, the same volume weighs 2400 tonnes. Similarly, 1 m³ of a typical topsoil with a bulk density of 1.0 will weigh 1 tonne (1000 kg) when dry and up to half as much again when moist.

### Soil structures

The pore space does not depend solely upon the size of the soil particles as shown in Figure 13.4 because they are normally grouped together. These **aggregates**, or peds, are groups of particles held together by the adhesive properties of clay and humus. The ideal arrangement of small and large pores is illustrated in Figure 13.7 alongside a dusty tilth with too few large pores and a cloddy tilth which has too many large pores.

A soil with a **simple structure** is one in which there is no observable aggregation. If this is because none of the soil particles are joined together, as in sands or loamy sands with low organic matter levels, it is described as **single grain** structure. Where all the particles are joined with no natural lines of weakness the structure is said to be **massive**. A **weakly-developed** structure is one in which aggregation is indistinct and the soil, when disturbed, breaks into very few whole aggregates but a lot of unaggregated material. This tends to occur in loamy sands and sandy loams. Soils with a high clay content form **strongly-developed** structures in which there are obvious lines of weakness and, when disturbed, aggregates fall away undamaged. The prismatic, angular blocky, round blocky, crumbs, and platy structures which are found in soils are illustrated in Figure 13.8.

### Development of soil structures

Soil structure develops as the result of the action on the soil components of **natural structure-forming agents**: wetting and drying, freezing and thawing, root growth, soil organisms and cultivations. The **wetting and drying cycle** can affect the whole rooting depth. Cracks open up in heavier soils as the clay shrinks. As well as being a major factor in the drying of deeper soil layers, the **roots** play an important part in soil structure by growing into the cracks and keeping them open. They help establish the natural fracture lines. In strong structures, a close-fitting arrangement of **prismatic** or **angular blocky** aggregates is readily seen. In soils with low clay content the roots are vital in maintaining an open structure.

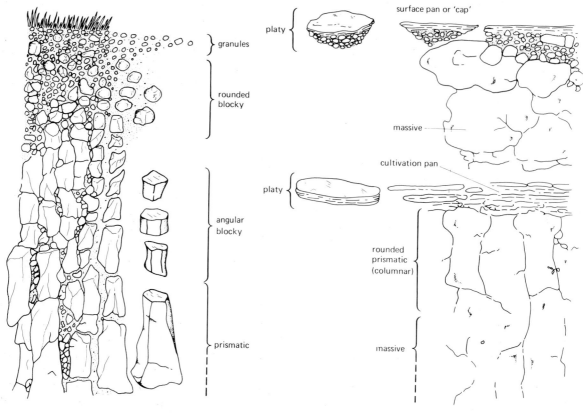

**Figure 13.8** *Soil structures*
*The soil profile on the left is composed of soil particles aggregated into structures which produce good growing conditions. Examples of structures which create a poor rooting environment are shown in the profile on the right.*

The exploring roots probe the soil, opening up channels where the soil is loose enough and producing sideways pressure as they grow. On death, the root leaves behind channels stabilized by its decomposed tissue for other roots to follow. Fine **granular** structures are developed under pastures by the action of the fibrous rooting over many years. The soil structure is greatly improved by the root ball and its physical influence is most easily appreciated by shaking out the soil crumbs from around the root of a tuft of well-established grass and comparing them with the structure of soil taken from a nearby bare patch. The shattering of clods by **frost action**, producing a frost mould, is largely confined to the surface layers and is vital in the management of clays. Freshly exposed land

is often referred to as **raw**; when weathered it becomes **mellow**. Once mellow, a seedbed is more easily prepared. The weathering process and influence of cultivation tend to produce rounded blocky **structures** and **rounded granules** in the cultivated zone (Figure 13.8).

Earthworms and other soil organisms play an important part in loosening soil, maintaining the network of drainage channels and stabilizing the soil structure. **Cultivation** of soil by hand or mechanized implements is undertaken to produce a suitable rooting environment for plants, to destroy pests and weeds and to mix in plant residues, manures and fertilizers. The use of cultivators can lead to the formation of platy layers or **pans** which are characterized by the lack of vertical cracks and form an obstruction to root or water movement. **Natural pans** develop in some soils as a result of fine material cementing a layer of soil together. In some sandy soils rich in iron oxide, these oxides cement together a layer of sand where there has been a fluctuating water table, to produce an iron pan. The collapse of surface crumb structure can lead to the formation of a **surface pan**, **crust** or 'cap'.

### Structural stability

**The structural stability of soil refers to its ability to resist deformation when wet.** Soil aggregates with little or no stability collapse spontaneously as they soak up water, i.e. they slake. Those high in fine sand or silt are particularly vulnerable to slaking. Aggregates with better stability maintain their shape when wetted for a short time but gradually pieces fall off if left immersed in water. Aggregates with **good structural stability** are able to resist damage when wet unless vigorously disturbed. Soils with a high clay content have better stability than those with low levels. Stability is also increased by the presence of calcium carbonate (chalk), iron oxides, and, most importantly, **humus**.

### Tilth

**The soil surface or seedbed should be carefully managed to produce the required tilth: the structure of the top 50 mm of the soil.** Sandy soils are easily broken down to the right size with cultivation equipment. Heavier soils are less easy to cultivate and benefit from weathering to produce a frost 'mould'.

The fineness of a seedbed should be related to the size of seeds but ideally consists of aggregate or crumbs between 0.5 and 5 mm in diameter (Figure 13.7). Cloddy surfaces lead to poor germination as well as poor results from soil herbicide treatments. Tilth is broken down by the rain on the soil surface. As soil crumbs break up, the particles fill in the gaps; this reduces infiltration rates. As the surface dries, a **cap** or **pan** is formed. Thus fine 'dusty' tilths should be avoided and the soil crumbs should be stable so that they can withstand the effect of rain until the crop is established. This is particularly important on fine sandy and silty soils which tend to have poor structural stability.

## CULTIVATIONS

In temperate areas the conventional preparation of land for planting is a thorough disturbance of the top 20–30 cm of soil. Residues of previous plantings and weeds are buried by digging or ploughing and with repeated passes of rakes or harrows a suitable **tilth** is created. This procedure is very demanding on energy, labour and time. Many of the cultivations tend to override the natural structure-forming agents and when undertaken at the wrong time they create pans or leave a bare, loose soil vulnerable to erosion. Some of the compaction problems are overcome by cultivating in **beds** which confines traffic to well-defined paths between the growing areas. The advent of effective herbicides has, in certain cases, enabled the inversion of soil to be eliminated. The use of powered implements has speeded up work and reduced the number of tillage passes. In some areas of horticulture the adoption of **minimum** or **zero tillage** has preserved natural structure while beneficially concentrating organic matter levels in the surface layers and reducing wind and water erosion.

### Ploughing and digging

Ploughing and digging are used to loosen and invert the soil. The land is broken up into clods and an increased area is exposed to weathering. As the soil is inverted weeds, plant residues and bulky manures are incorporated. The depth of ploughing or digging should be related to the depth of topsoil because bringing up the subsoil reduces fertility in the vital top layers, seriously affecting germination of seeds and establishment of plants. If deeper layers are to be loosened a **subsoiler** should be used. In plastic soil conditions the plough can smear the soils, particularly when the wheels of the tractor spin in the furrow bottom. These **plough pans** tend to develop with successive ploughing to the same depth. Their incidence can be reduced by ploughing at different depths or attaching a subsoil tine. Digging with a spade does not produce a cultivation pan and is still used on small areas to break this type of pan. **Spading machines** or **rotary diggers** imitate the digging action without the disadvantages of ploughing but tend to be very slow.

### Rotary cultivators

Rotary cultivators are used to create a tilth on uncultivated or on roughly prepared ground. The type of tilth produced depends on suitable adjustment of forward speed, rotor speed, blade design and layout, shield angle and depth of working. The 'hoe' blade is normally used for seedbed production but does have the disadvantage of smearing plastic soils at the cultivation depth, producing a **rotovation pan**. 'Pick' tines produce a rougher tilth but less readily cause a pan. Subsoil tines can be fitted to prevent these pans developing.

### Harrowing and raking

Harrowing and raking are methods of levelling soils, incorporating fertilizers and producing a suitable tilth on the roughly prepared ground. The soil must be in a **friable** condition for this operation and it is made easier if the top layers have been suitably weathered. The impact of the tines breaks the clods.

'Progressive' type cultivators were introduced essentially to loosen coarse structured clays by drawing through the soil banks of tines, increasing in depth from the front, to cultivate the soil from the top down in one pass. Although this requires powerful tractors to pull, especially if subsoiling tines are attached, it is a time-saving operation and the recompaction inherent in multiple pass methods is reduced. These cultivators should not be used on well-structured soils where full depth loosening is unnecessary.

'Under-loosening' cultivators have been designed to loosen compact topsoils without disturbing the surface, which ensures a level, clod-free and organic-matter-rich tilth. Used under the right conditions these implements improve water movement and crop growth. However, loosened soils are more susceptible to compaction and consequently the equipment should only be used when compaction is known to be present.

### Subsoiling

Subsoiling is used to improve soil structure below plough depth by drawing a heavy cultivation tine through the soil to establish a system of deep cracks in compacted zones. This helps the downward movement of water, circulation of air and penetration of roots (*see* Figure 13.9). The operation is most effective when the subsoil is friable and the surface dry enough to be able to withstand the heavy tractor that is needed (*see* loadbearing). Effective subsoiling is made easier if the top surface is loosened by prior cultivation. Although the draught is higher, subsoil disturbance is increased substantially by attaching inclined blades or wings. Successful subsoiling is accompanied by a lift in the soil surface, **soil heave**, which usually makes it unsuitable for improving conditions in playing fields.

Subsoiling should only be used when the cause of any waterlogging is related to a soil structure fault (*see* drainage). Slow subsoil permeability caused by high clay content is usually rectified with mole drainage but if the soil is too sandy or stony a subsoiler can be used so long as the cracks created lead the water into a natural or artificial drainage system. Massive structures and soil pans created by machinery are readily burst by sub-

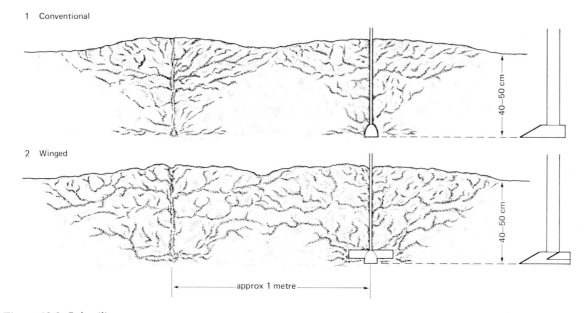

1  Conventional

2  Winged

approx 1 metre

40–50 cm

40–50 cm

**Figure 13.9** *Subsoiling*
*The subsoiler is drawn through the soil to burst open compacted zones. It leaves cracks which remain open to improve aeration, drainage, and root penetration. The cracks created should link up with artificial drainage systems unless the lower layers are naturally free draining.*

soilers used in the right conditions but some natural pans are too strong for normal farm equipment.

The problem of cultivation pans can be dealt with by using conventional subsoilers or by attaching small subsoil tines to the cultivation equipment. This tends to increase the power requirement but eliminates the pan as it is created.

## MANAGEMENT OF MAIN SOIL TYPES

### Sandy soils

Sandy soils are usually considered to be easily cultivated but serious problems can occur because the particles readily pack together, especially when organic matter levels are low. Consequently many sandy soils are difficult to firm adequately without causing over-compaction. Pans near the surface caused by traffic and deeper cultivation pans frequently occur on sandy soils, resulting in

reduced rooting and water movement. **Subsoiling** is frequently undertaken on a routine basis every 4–6 years although the need can be reduced by keeping machinery off land when it has low **load-bearing** strength and by encouraging natural structure-forming agents.

Coarse sands have low water-holding capacity which makes them vulnerable to drought, particularly in drier areas. This is not such a disadvantage if irrigation equipment is installed and water is readily obtainable. In many categories of horticulture there is a demand for soils with **good workability**. Coarse sands, loamy sands and sandy loams have the advantage of good porosity and can be cultivated at field capacity. Sands tend to go acid rapidly and are vulnerable to overliming because of their **low buffering capacity**.

### Silts and fine sands

These can be very productive soils because of

their good water-holding capacity and, while organic matter levels are kept above 4 per cent, their ease of working. However silts and fine sands present soil management problems, especially when used for intensive plantings, because they have **weak structure**, are vulnerable to **surface capping**, and are easily compacted to form **massive structures**. To achieve their high potential, efficient drainage is vital to maximize the rooting depth. Fine tilths in the open should be avoided, especially in autumn and early spring, because frosts and heavy rainfall reduce the size of surface crumbs. For the same reason, care should be taken with irrigation droplet size which, if too large, can damage the surface structure. Improving soil structure is not easy after winter root crop harvesting or orchard spraying on wet soils because low clay content results in very little cracking during subsequent wetting and drying cycles. Improvement therefore depends on other natural structure forming agents or on subsoiling.

## Clay soils

Clay soils tend to be **slow draining**, **slow to warm** up in spring, and have **poor working properties** (*see* workability). A serious limitation is that the soil is still plastic **at field capacity**, which delays soil preparation until it has dried by evaporation. Permanent plantings are established to avoid the need to rework the soil. Playing surfaces created over clays have severe limitations, particularly when required for use in all weather conditions. Where high standards have to be maintained, as in golf greens, fine turf is established in a suitable growing medium overlying the original soil. However, a high clay content is an advantage for the preparation of cricket squares where a hard, even surface is required but is played on only in drier weather. Increasingly, heavily used areas are replaced by artificial surfaces.

Horticultural cropping of clays is limited to summer cabbage, Brussels sprouts and to some top fruit in areas where the water table does not restrict rooting depth. Under drainage is normally necessary. In wetter areas most clays are put down to grass. Timeliness, encouraging the annual drying cycle of the soil profile, and maximizing the effect of weathering to help cultivations are essential for successful management of clay soils.

## Peat soils

Peat soils have very many advantages over mineral soils for intensive vegetable and outdoor flower production. Fenland soils and Lancashire Moss of England; peatlands of the midland counties of Ireland; the 'muck' soils of North America; and similar soils in the Netherlands, Germany, Poland and Russia have proved valuable when their limitations to commercial cropping have been overcome.

Well-drained peat at the correct pH is an **excellent root environment**. It has a very much higher water-holding capacity than the same volume of soil and yet gaseous exchange is good. Root development is uninhibited because friable peat offers hardly any mechanical resistance to root penetration. This leads to high quality root crops which are easily cleaned. These cultivated peat lands **warm up quickly** at the surface because the sun's energy is efficiently absorbed by their dark colour, with consequent rapid crop growth. These soils have a very **low power requirement** for cultivation, are free of stones, and can be worked over a wide moisture range.

Plant nutrition is complicated by natural **trace element deficiencies** and the effect of pH on plant **nutrient availability**. Peat has **poor load-bearing** characteristics and specialized equipment is often needed to harvest in wet conditions. Whilst peat warms up quickly on sunny days, its dark surface makes it vulnerable to air frost because it acts as an efficient radiator. This is combated by firming the surface and keeping it moist. Weeds grow well and their control is made more difficult by the ability of peat to absorb and **neutralize soil acting herbicides**. The high organic matter levels also make the peats and sandy peats **vulnerable to wind erosion** in spring when the surface dries out and there is no crop canopy to protect it.

## PRACTICAL EXERCISES

1 Examine the **root systems** of different plants in a range of soils. Dig holes alongside a growing plant and wash away the soil from the roots, look at the root system and note how the soil structure affects root growth and vice versa.
2 Examine different rock faces for evidence of **weathering**. Look at the base of vertical rock faces for evidence of rock debris. Account for any differences observed. Note the huge accumulations of debris called scree on exposed mountain sides.
3 Walk back from a sea-shore and examine the shingle, sand, or other rock fragments from the beach to the recognizable soil. Look for **plant colonization** and the development of a humus-rich topsoil.
4 Determine the texture of a range of soils by the following rough mechanical analysis:
   (a) Dry soil and then rub it through a 2 mm sieve to remove stones. Heat a sample strongly on a metal tray for an hour to burn off organic matter (*see* Chapter 15, Exercise 4).
   (b) Place 100 g prepared soil in a 200 g coffee jar and add a teaspoonful of Calgon and a pinch of sodium carbonate. One-third fill Jar No. 1 with water. Leave to stand overnight then shake vigorously for as long as possible to separate all the particles.
   (c) Add water until 10 cm deep. Shake very vigorously and leave soil to settle for 45 seconds then quickly and steadily pour into Jar No. 2 the water which contains silt and clay, leaving the deposited sand in Jar No. 1.
   (d) Top up water in Jar No. 1 to 10 cm, shake, stand for 45 seconds, decant into Jar No. 3 again leaving the sand in Jar No. 1. Repeat until clean water decanted off the sand. Dry and weigh sand. Record percentage of sand.
   (e) Top up all jars containing silt and clay to 10 cm mark. Shake all jars vigorously and leave to stand for 8 hours. Pour off water containing clay and keep deposited silt. Dry silt and weigh. Record percentage of silt.
   (f) Plot percentage sand and percentage silt on Textural Triangle (*see* Figure 13.6) and identify soil type.

5 Determine **texture** of soil by feel and compare results from mechanical analysis method (*see* texturing by feel).
6 Make up a paste of water and sand. Smear thickly over a white tile and allow to dry. Repeat using silt, clay, and clay mixed thoroughly with lime. Note the differences in cracking on drying.
7 Dig soil profile pits in different localities and identify **soil structures**. Examine faces of soil exposed in trenches and building excavations to observe soil down to its parent material.
8 Determine the **bulk density** of soils before and after cultivation and at different depths in the soil. Repeat on different soils and on uncultivated land.

## FURTHER READING

Bradshaw, M. *The Earth: Past, Present and Future. An Introduction to Geology* (Hodder & Stoughton, 1980).

Coker, E.G. *Horticultural Science and Soils* (Macmillan, 1970).

Curtis, L.F., Courtney, F.M., and Tredgill, S. *Soils in the British Isles* (Longman, 1976).

Handreck, K., and Black, N. *Growing Media for Ornamental Plants and Turf* (New South Wales University Press, 1989).

Mackney, D. (Ed.) *Soil Type and Land Capability*. Soil Survey Technical Monograph No. 4 (HMSO, 1974).

Simpson, K. *Soil* (Longmans Handbooks in Agriculture, 1983).

White, R.E. *Introduction to the Principles and Practice of Soil Sciences* (Blackwell Scientific Publications, 1979).

# 14

# Soil Water

*Plants require a constant supply of water to maintain growth. Water is taken up by the plant with minerals by the roots and is lost by* **transpiration**. *Approximately 500 kg of water is needed to replace the water lost from the leaves over a period when the plant grows by 1 kg of dry weight.*

*In this chapter the characteristics of soil water are established whilst following the process of wetting a dry growing medium, then the drying out of a fully wetted one. The concept of available water is related to the ability of the plant root to extract water from different growing media with different moisture contents. Irrigation and drainage are explained as the means of maintaining optimum water levels in the soil. The chapter is concluded by an examination of the nature of water quality and water conservation.*

## WETTING OF A DRY SOIL

**Rainfall** is recorded with a raingauge (*see* Figure 14.1) and is measured in millimetres of water. Thus '1 mm of rain' is the amount of water covering any area to a depth of 1 mm. Therefore '1 mm of rain' on one hectare of land is equivalent to $10 \, m^3$ or $10\,000$ litres of water per hectare.

As rain falls on a dry surface the water either soaks in (**infiltration**) or runs off over the surface as **surface run-off**. **Ponding** is the accumulation of water on the surface as a result of infiltration rates slower than rainfall. Ponding leads to capping which further reduces infiltration rates. Soil surfaces can be protected with mulches and care should be taken with water application rates during irrigation.

### Saturated soils

**As water soaks into the dry soil the surface layers become saturated or waterlogged, i.e. water fills all the pore spaces.** As water continues to enter the soil it moves steadily downwards, with a sharp boundary between the saturated zone and dry, air-filled layers, as shown in Figure 14.2. So long as water continues to soak into the soil, this wetting front moves to greater depths. When rainfall ceases the water in the larger soil pores continues to move downwards. The water which is removed from wet soil by gravity is known as **gravitational water**.

**Water is held in the soil in the form of water films around all the soil particles and aggregates. Forces in the surface of the water films, surface tension, hold water to the soil particles against the forces of gravity and the suction force of roots.** As the volume of water decreases, its surface area and hence its surface tension becomes proportionally greater until, in very thin films of

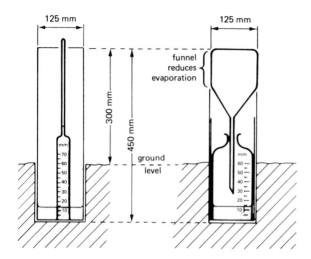

**Figure 14.1** *Raingauges*
*A simple raingauge consists of a straight-sided can in which the depth of water accumulated each day can be measured with a dip stick. An improved design incorporates a funnel, to reduce evaporation, and a calibrated collection bottle. A raingauge should be set firmly in the soil away from overhanging trees etc. and the rim should be 300 mm above ground to prevent water flowing or bouncing in from surrounding ground.*

water, it prevents the reduced volume of water from being removed by gravity. A useful comparison can be seen when one's hands are lifted from a bowl of water. They drip until the forces in the

surface of the thin film become equal to the forces of gravity acting on the remaining small volume of water over the hands.

**Field capacity (FC)**

**A soil that has been saturated, then allowed to drain freely without evaporation until drainage effectively ceases, is said to be at field capacity. This may take two days or less on sandy soils, but far longer on clay where the process of drainage may continue indefinitely.**

At field capacity the soil pores less than about 0.1 mm remain full of water, whereas in the **macropores** air replaces the gravitational water as illustrated in Figure 14.2 (*see* porosity).

The amount of water held at field capacity is known as the **moisture holding capacity** (MHC) or **water holding capacity** (WHC). Examples are given in Table 14.1. The MHC is expressed in millimetres of water for a given depth of soil. Thus a silty loam soil 300 mm deep holds 65 mm of water when at field capacity. Conversely, if a silty loam had become completely dry to 300 mm depth, it would require 65 mm of rain or irrigation water to return it to field capacity. Since 1 mm of water is equivalent to $10 \, m^3$/ha, a hectare of silty loam would hold $650 \, m^3$ water in the top 300 mm

**Table 14.1** Soil water holding capacity
The amount of water in a given depth of soil at field capacity can be calculated by simple proportion.

| Soil texture | Water held in 300 mm soil depth (mm) | | |
| --- | --- | --- | --- |
| | at field capacity i.e. moisture holding capacity | at permanent wilting point | Available water |
| Coarse sand | 26 | 1 | 25 |
| Fine sand | 65 | 5 | 60 |
| Coarse sandy loam | 42 | 2 | 40 |
| Fine sandy loam | 65 | 5 | 60 |
| Silty loam | 65 | 5 | 60 |
| Clay loam | 65 | 10 | 55 |
| Clay | 65 | 15 | 50 |
| Peat | 120 | 30 | 90 |

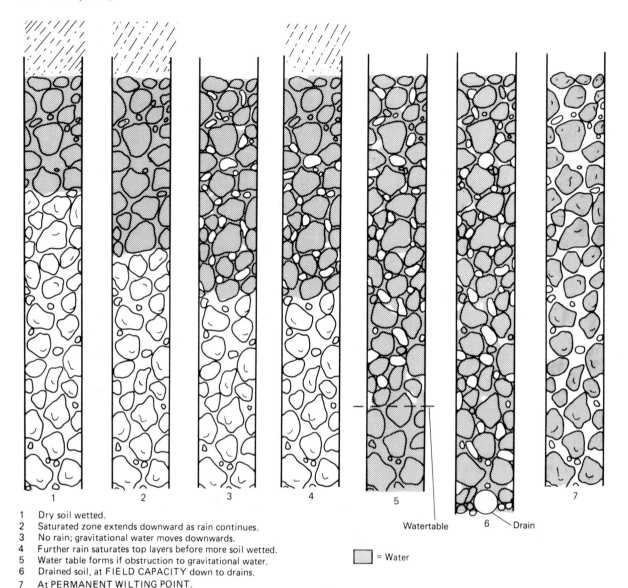

1   Dry soil wetted.
2   Saturated zone extends downward as rain continues.
3   No rain; gravitational water moves downwards.
4   Further rain saturates top layers before more soil wetted.
5   Water table forms if obstruction to gravitational water.
6   Drained soil, at FIELD CAPACITY down to drains.
7   At PERMANENT WILTING POINT.

Watertable

□ = Water

**Figure 14.2** *Water in the soil.*

when at field capacity (*see* rainfall). The principle described enables water-holding capacity or irrigation requirement to be determined for any soil depth. The amount of water required to return a soil to field capacity is called the soil moisture deficit.

## Watertables

**Groundwater** occurs where the soil and underlying parent material are saturated; the **watertable** marks the top of this saturated zone, which fluctuates over the seasons, normally being much higher in winter. In wetlands the watertable is very near the soil surface and the land is not suitable for horticulture until the watertable of the whole area is lowered (*see* drainage).

Where water flows down the soil profile and is impeded by an impermeable layer such as saturated clay or silty clay, a **perched watertable** is formed. Water from above cannot drain through the impermeable barrier and so a saturated zone builds up over it. **Springs** appear at a point on the landscape where an overlying porous material meets an impermeable layer at the soil surface, e.g. where chalk slopes or gravel mounds overlie clay.

### Capillary rise

**Capillary rise occurs only from saturated soils. Water is drawn upwards from the watertable in a continuous network of pores.** The height to which water will rise depends on the continuity of pores, and on their diameter. In practice the rise from the watertable is rarely more than 20 mm for coarse sands, typically 150 mm in finer textured soils but substantially greater in silty soils and on chalk.

The upward movement of water in these very fine pores is very slow. The principle of capillary rise is used in watering plants grown in containers (*see* capillary benches). Several 'self-watering' containers also depend on capillary rise from a water store in their base (*see* aggregate culture).

## DRYING OF A WET SOIL

**Soil water is lost from the soil surface by evaporation and from the rooting zone by plant transpiration.**

### Evaporation

The rate of water loss from the soil by **evaporation** depends on the drying capacity of the atmosphere just above the ground and the water content in the surface layers. The evaporation rate is directly related to the **net radiation** from the sun which can be measured with a solarimeter. Evaporation rates increase with higher air temperatures and wind speed or lower humidity levels. As water evaporates from the surface, the water films on the soil particles become thinner. The surface tension forces in the film surface become proportionally greater as the water volume of the film decreases. This leaves water films on the particles at the surface with a high surface tension compared with those in the films on particles lower down in the soil. Water moves slowly upwards to restore the equilibrium. **Whilst the surface layers are kept moist by water moving slowly up from below, the losses by evaporation in contrast are quite rapid. Consequently the surface layers can become dry and the evaporation rate drops significantly after '5 mm of water' is lost. Evaporation virtually ceases after the removal of '20 mm of water' from the soil. A dry layer on the soil surface helps conserve moisture in the lower layers.** Water loss from the soil surface can also be reduced by mulches. The evaporation from the soil surface is almost eliminated by a leaf canopy which shades the surface, reduces air flow and maintains a humid atmosphere over the soil.

### Evapotranspiration

As a soil becomes covered by a leaf canopy the rate of water loss becomes more closely related to transpiration rates. **The potential transpiration rate represents the estimated loss of water from plants grown in moist soil with a full leaf canopy. It can be calculated from weather data (*see* Table 14.2).**

As water is removed by roots it is slowly replaced by the water film equilibrium, but rapid water uptake by plants necessitates root growth towards a water supply. At any point when water loss exceeds uptake, the plant loses turgor and may wilt. This tends to happen in very drying conditions even when the growing medium is moist. Wilting is accompanied by a reduction in

**Table 14.2**   Potential transpiration rates

The calculated water loss (mm) from a crop grown in moist soil with a full leaf canopy, over different periods of time and based on weather data collected in nine areas in the British Isles.

| Area | April | May | June | July | Aug. | Sept. | Summer | Winter | Annual |
|------|-------|-----|------|------|------|-------|--------|--------|--------|
| Ayr | 46 | 81 | 90 | 83 | 65 | 38 | 403 | 71 | 474 |
| Bedford | 50 | 78 | 89 | 91 | 80 | 43 | 430 | 70 | 500 |
| Cheshire | 53 | 75 | 83 | 88 | 76 | 44 | 420 | 104 | 500 |
| Channel Isles | 51 | 86 | 91 | 99 | 84 | 46 | 457 | 104 | 561 |
| Essex | 50 | 79 | 98 | 98 | 83 | 45 | 450 | 80 | 530 |
| Hertfordshire | 49 | 79 | 91 | 94 | 80 | 43 | 435 | 75 | 510 |
| Kent | 50 | 79 | 93 | 96 | 83 | 44 | 440 | 65 | 505 |
| Northumberland | 44 | 64 | 81 | 76 | 60 | 34 | 359 | 70 | 429 |
| Pembroke (Dyfed) | 46 | 75 | 84 | 81 | 74 | 44 | 404 | 104 | 508 |

carbon dioxide movement into the leaf which in turn reduces the plant's growth rate (*see* photosynthesis). The plant recovers from this **temporary wilt** as the rate of water loss falls below that of the uptake, which usually occurs in the cool of the evening onwards. **Continued loss of water causes the plants to reach the permanent wilting point which is the point when they wilt but do not regain turgor overnight because their roots can extract no more water within the rooting zone.** When the soil has reached the **permanent wilting point** (PWP) there is still water in the smallest of soil pores, within clay particles, and in combination with other soil constituents, but it is too tightly held to be removed by roots. Typical water contents of different types of soil at their PWP are given in Table 14.1.

**Available water**

Roots are able to remove water held at tensions up to 15 atmospheres within the rooting zone, and gravitational water drains away. **Consequently the available water for plants is the moisture in the rooting depth between field capacity and the permanent wilting point. The available water content (AWC) of different soil textures is given in Table 14.1.** Fine sands have very high available water reserves because they hold large quantitites of

water at FC and there is very little water left in the soil at PWP (*see* sands). Clays have lower available water reserves because a large proportion of the water they hold is held too tightly for roots to extract (*see* clay).

Roots remove the water from films at field capacity very easily. Even so, plants can wilt temporarily and any restriction of rooting makes wilting more likely. Water uptake is also reduced by high soluble salt concentrations and by the effect of some pests and diseases (*see* vascular wilt diseases). As the soil dries out, the water films become thinner and the water more difficult for the roots to extract. After about half the available water content has been removed, wilting becomes significantly more frequent. Slowing of growth rates is minimized by irrigating when available water levels fall below this point. Plants grown under glass are often irrigated more frequently to maintain a suitably formulated growing medium near to field capacity. This ensures maximum growth rates since the roots have access to 'easy' water, i.e. water removed by low suction force.

**Workability of soils**

**The number of days each year that are available for soil cultivation depends on the weather but more specifically on soil consistency. This describes**

the effect of water on those physical properties of the soil influencing the timing and effect of cultivations. It is assessed in the field by prodding and handling the soil. A very wet soil can lose its structure and flow like a thick **fluid**. In this state it has no **load-bearing strength** to support machinery. As the soil dries out the soil become **plastic**; the particles adhere and are readily moulded. In general, the soil is difficult to work in this condition because it sticks to the cultivation equipment, has insufficient load-bearing strength, is readily compacted, and is easily smeared by cultivating equipment. As the soil dries further it becomes **friable**. At this stage the soil is in the ideal state for cultivation because it has adequate load bearing strength but the soil aggregates readily crumble. If the soil dries out further to a harsh consistency the load bearing strength improves considerably, but whilst coarse sands and loams still readily crumble in this condition, soils with high clay, silt, or fine sand content form hard resistant clods. The friable range can be extended by adding organic matter (*see* humus). At a time when bulky organic matter is more difficult to obtain it is important to note that a fall in soil humus content narrows the friable range. This allows less latitude in the timing of cultivations and increases the chances of cultivations being undertaken when they damage the soil structure. **Timeliness is the cultivation of the soil when it is at the right consistency. Whereas many sands and silts can be cultivated at field capacity, clays and clay loams do not become friable until they have dried out to well below field capacity.**

## DRAINAGE

**Drainage is the removal of gravitational water from the soil profile. As this water leaves the macropores, air that takes its place enables gaseous exchange to continue.** Horticultural soils should return to at least 10% air capacity in the top half metre within one day of being saturated (*see* porosity). Some soils, notably those over chalk or gravel, are naturally free draining but many have underlying materials which are impermeable or only slowly permeable to water. In such cases artificial drainage, sometimes referred to as field drainage or under drainage, is put in to carry away the gravitational water. This helps the soil to reach its field capacity rapidly but does not reduce its moisture holding capacity.

**Well-drained soils** are those which are rarely saturated within the upper 900 mm except during or immediately after heavy rain. Uniform brown, red, or yellow colours indicate an **aerobic** soil, i.e. a soil in which oxygen is available. **Imperfectly-drained soils** are those which are saturated in the top 600 mm for several months each year. These soils tend to have less bright colours than well-drained soil. Greyish or ochreous colours are distinct at 450 mm, giving a characteristic rusty mottled appearance. **Poorly-drained soils** are saturated within the upper 600 mm for at least half the year and are predominantly grey.

### Symptoms of poor drainage

These include restricted rooting; reduced working days for cultivation; weed, pest and disease problems; and excess fertilizer requirements. Soil pits dug in appropriate places reveal the extent of the drainage problem and help pinpoint the cause which is the basis of finding the solution. The current watertable is indicated by the level of water that develops in the pit. Further indications of poor drainage are the presence of high organic matter levels (*see* organic soils) and small black nodules of manganese dioxide. Topsoils waterlogged for long periods in warm conditions have a smell of bad eggs (*see* sulphur cycle). The presence of compacted zones should be looked for as they indicate obstructions to the flow of gravitational water (*see* bulk density).

Soil colours show the history of waterlogging in the soil. Whereas free drainage is indicated by uniform red, brown or yellow soil throughout the subsoil, the iron oxide which gives soils these colours in the presence of oxygen is reduced to grey or blue forms in **anaerobic** conditions, i.e. when no oxygen is present. Zones of soil which are saturated for prolonged periods have a dull grey appearance, referred to as **gleying**. Reliance on colour alone as an indication of drainage conditions is not recommended because it persists for a long time after efficient drainage is established.

**Structural damage** whether caused by water (*see* stability), machinery (*see* cultivation), or by accumulations of iron (*see* natural pans) is an obstruction to water flow in the soil profile. Platy structures near the surface can be broken with cultivating equipment on arable land or spiking on grassland; but those deeper in the soil should be burst by subsoiling. If water cannot soak away from well-structured rooting zones, artificial drainage is required.

**Low permeability**

Low permeability of subsoils is the major reason for under-drainage in horticultural soils (*see* porosity). Clay, clay loam and silty clays when wetted become almost impermeable as the clay swells and the cracks close. Lines of pipes are placed in the subsoil to intercept the trapped water. An even **gradient** from their highest point to the outfall in a ditch should be established to prevent silting up and **silt traps** should be placed at regular intervals to help to service the system at points where there is a change of gradient or direction (*see* Figure 14.3). The spacing between the lines of pipes depends on the permeability of the soil, a maximum of 5 metre intervals being necessary in clay subsoils. Soil permeability and the land use dictates the **depth of the drains** which is normally more than 60 cm. Drains should be set deeply in cultivated land where heavy equipment and deep cultivation might disturb the pipes. Shallow drains can be used where rapid drainage is a high priority and the pipes are not likely to be crushed by heavy vehicles or severed by cultivating equipment, e.g. sports grounds. **The diameter of the pipe** depends on the gradient available and the amount of water to be carried when wet conditions

(a) *Simple Interceptor Drainage System:*

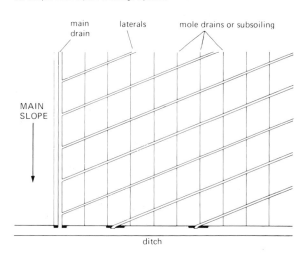

(b) *Outfalls*

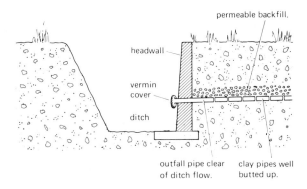

(c) *Silt trap/Inspection Chamber*

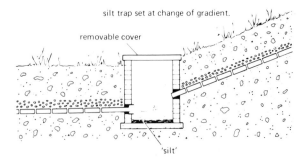

**Figure 14.3** *Drainage.*

prevail. Most frequently used are 75 mm and 100 mm diameter pipes, usually leading into larger drains.

Pipes are made of clay or plastic. **Tiles** (clayware) are usually 300 mm sections of pipe butted tightly together to allow entry of water but not soil particles. It is recommended that they are covered with permeable fill, usually stones or clinker, to improve water movement into the drains. **Plastic pipes** usually consist of very long lengths of pipe perforated by many small holes.

Water has to be discharged into a ditch above the wet season water level to allow water out of the pipes. The outlet of the drains is very vulnerable to damage and so it should consist of a strong, long pipe set flush in a concrete or brick **headwall** so that it is not dislodged by erosion in the ditch nor by people using it as a foothold. Pipes not set flush should be glazed to prevent frost damage. **Vermin traps** should be fitted to prevent pipes being blocked by nests or dead animals.

**Mole drainage** is very much cheaper than pipe drainage. A mole plough draws a 75 mm bullet followed by a 100 mm plug through the soil at a depth of 500–750 mm from a ditch up the slope of a field or across a pipe drain system with permeable backfill (*see* Figure 14.3). The soil should be plastic at the working depth so that a tunnel to carry water is created. The soil above should be drier so that some cracks are produced as the implement is drawn slowly along. These cracks improve the soil structure and conduct water to the mole drain. Sandy and stony areas are unsuitable because tunnels are not properly formed or collapse as water flows. Tunnels drawn in clay soils can remain useful for 10–15 years but in wetter areas their useful life may be nearer 5 or even as little as 2 years.

Pipe drainage is usually combined with secondary treatments such as mole drainage or subsoiling to achieve effective drainage at reasonable costs. Deep subsoiling improves soil permeability and the pipes carry the water away. Installation costs can be reduced because pipes can be laid further apart. Similarly mole drainage over and at

right angles to the pipes enables them to be spaced 50–100 metres apart.

**Sandslitting** is used on sports grounds to remove water from the surface as quickly as possible. It involves cutting trenches at frequent intervals in the soil and infilling with carefully graded sand that conducts water from surface to a free draining zone under the playing surface.

Water that spills out on to lower ground from springs can be intercepted with **ditches** placed at the junction of the permeable and impermeable layers. French drains can be placed around impermeable surfaces such as concrete hardstandings and stone patios to intercept the run-off.

### Maintenance of drainage systems

Under-drainage is very expensive to install and must be serviced to ensure that the investment is not wasted. **Ditches** need regular attention because they are open to the elements. Weed growth should be controlled, rubbish cleared out and collapsed banks repaired, because obstructions lead to silting up or the undercutting of the bank. The design of the ditches depends on the soil type and should be maintained when being repaired. **Drain outlets** are a particularly vulnerable part of the drainage system, especially if not set into a headwall. They should be marked with a stake (holly trees were traditionally used in some areas) and inspected regularly after the soil returns to field capacity. Blockages should be cleared with rods and vermin traps refitted where appropriate. Wet patches in the field indicate where a blockage in a pipe has occurred. The pipe should be exposed and the cause of the obstruction remedied. Silted-up pipes can be rodded, broken sections replaced, or dislodged pipes realigned. **Silt traps** need to be cleaned out regularly to prevent accumulated soil being carried into the pipes.

At all times it should be remembered that the drains only carry away water which reaches the pipes. Every effort must be made to maintain good soil permeability and to avoid compaction problems. Subsoilers should be used to remedy subsoil structural problems. Once drainage has been installed the soil dries more quickly, leading to better soil structure because cracks appear more extensively and for longer periods. Deeper layers of the soil are dried out as roots explore the improved root environment, which gives another turn in the improving cycle.

## IRRIGATION

**Irrigation is used to prevent plant growth being limited by water shortage. Irrigation should be seen as a husbandry aid: an addition to otherwise sound practice.** It is assumed in the following that water is being added to a well-drained soil unaffected by capillary rise. The need for irrigation depends on available water in the rooting zone and the effect of water stress on the plant's stage of growth. **Response periods** are the growth stages when the use of irrigation during periods of rainfall deficiency is likely to show economic benefits. In general all plants benefit from moist seedbeds and vegetative growth is maximized by eliminating water stress. Initiation of flowering and fruiting is favoured by drier conditions. The response periods of a range of plants grown in the UK is given in Table 14.3.

### Irrigation plan

**In general terms water is added to a soil when moisture levels fall to 50% of available water content in the rooting zone. Outdoors 25 mm of water is the minimum that should be added at any one time in order to reduce the frequency of irrigation, to reduce water loss by evaporation and to prevent the development of shallow rooting.** On most soils the amount of water added should be such as to return the soil to field capacity. Addition of water to clays and clay loams should be minimized so as not to reduce the vital drying and wetting cycles and if they have to be irrigated they should not be returned to field capacity in case rain follows (*see* ponding). Irrigation should never

result in fertilizers being leached from the rooting depth unless it is the specific objective, as in flooding.

Most recommendations are given in a simplified form taking the above points into account. The recommended plan is usually expressed in terms of how much water to apply, at a given soil moisture deficit, for a named crop on a soil of stated available water content. Thus for outdoor grown summer lettuce crop grown on soils of a medium available water content, 25 mm of water should be added when a 25 mm soil moisture deficit occurs. This would require the application of 250 000 litres per hectare or 25 litres per square metre. Further examples are given in Table 14.3.

### Soil moisture deficits

**Soil moisture deficit (SMD) is the amount of water required to return the growing medium to field capacity.** SMD can be calculated by keeping a **soil water balance sheet**. The account is conveniently started after rain returns the soil to field capacity, i.e. when SMD is zero. In Britain it is assumed that, unless it has been a dry winter, the soil is at field capacity until the end of March. From the first day of April a day-by-day check can be made of water gains and losses. **Rainfall** varies greatly from year to year from one locality to the next and so it should be determined on site (*see* raingauge) or obtained from a local weather station. Water loss for each month does not vary very much over the years and so potential transpiration rates based on past records can be used in the calculation. There are potential transpiration rate figures available for all localities having weather stations. Examples are given in Table 14.2. These figures can be used when calculating water loss but until there is 20% leaf canopy a maximum SMD of 20 mm is not exceeded because in the early stages water loss is predominantly from the soil surface by **evaporation**. A worked example is given in Table 14.4.

In protected cropping all the water that plants require has to be supplied by the grower, who

**Table 14.3**  Irrigation guide

| Plants | Response periods Growth stages at which to water and time when they occur | | Irrigation plan (mm of water at mm SMD) | | |
| --- | --- | --- | --- | --- | --- |
| | | | A low AWC[1] | B medium AWC | C high AWC[2] |
| *Beans*, runner | Early flowering onwards | June to August | 25 at 25+ | 50 at 50+ | 50 at 75 |
| *Brussels sprouts* | When lower buttons are 15–18 mm diameter | August to October | 40 at 40 | 40 at 40 | 40 at 40 |
| *Carrots* | Throughout life | June to September | 25 at 25 | 40 at 50 | |
| *Cauliflowers* early summer | Throughout life | April to June | 25 at 25 | 25 at 25 | |
| *Flowers* annual and biennial | Throughout life between April and early September | April to September | 25 at 25 | 25 at 25 | 25 at 25 |
| perennial | Throughout life | April to September | 25 at 25 | 50 at 75 | 50 at 75 |
| *Lettuce* summer | Throughout life | April to August | 25 at 25 | 25 at 25 | 25 at 50 |
| *Nursery stock* trees and shrubs | a. to establish newly planted stock | April to June | 25 at 25 | 25 at 50 | 25 at 50+ |
| | b. established stock | May to July | 25 at 25 | 25 at 50 | 25 at 50+ |
| | c. to aid early lifting | September | 25 at 25+ | 25 at 50+ | 25 at 50+ |
| *Potatoes*, first early | After tuberization reaches 10 mm diameter | May to June | 25 at 25–35 | 25 at 50–35 | 25 at 50–35 |
| maincrop and second early | From time tubers reach marble stage onwards | June to August | 25 at 25 | 25 at 50–40 | 25 at 25–40 |
| *Rhubarb* | When pulling has stopped | May to September | 40 at 50+ | 40 at 50+ | 50 at 75+ |
| *Strawberries* runners at planting | Date of planting to September | | 50 at 50 | 50 at 75 | 50 at 75 |
| fruiting | May to June September to October | | 50 at 50 | 50 at 75 | 50 at 75 |
| *Top fruit* Apples, heavy cropping and mature | July to September | | When SMD is more than 50 mm apply 50 mm of water to suffice for two weeks. Then continue irrigation to make the total water supply (rainfall + irrigation) equal to 50 mm/ fortnight for the remainder of July, 40 mm/fortnight in August and 25 mm/fortnight in September | | |
| Pears | July to August | | | | |

[1] less than 40 mm available water per 300 mm soil, e.g. gravels, coarse sands.
[2] more than 65 mm available water per 300 mm soil, e.g. silts, peats.

must therefore have complete control over irrigation. With experience the grower can determine water requirements by examining plants, soil, root balls, or by tapping pots. A **tensiometer** can be used to indicate the soil water tension but while this is useful to indicate when to water, it does not show directly how much water is needed. **Evaporimeters** distributed through the crop can give the water requirement by showing how much water has been evaporated. A **solarimeter** measures the total radiation received from the sun and the readings obtained can be used to calculate the crop water losses, often expressed in litres per square metre for convenience.

**Methods of applying water**

These should be carefully related to crop requirements, climate and soil type. On a small scale, **watering cans** or **hoses** fitted with trigger lances can be used but care should be taken to avoid damaging the structure of the growing medium. Water can be sprayed from fixed or mobile equip-

**Table 14.4**  Soil water balance sheet for established nursery stock grown on sandy loam (AWC 55 mm per 300 mm) in Essex. The irrigation plan is to apply 25 mm water if a 50 mm SMD is reached (*see* Table 14.3)

| Week beginning: | Water loss[1] (mm) | Water gain[2] Rainfall (mm) | Irrigation (mm) | SMD at end of week[3] (mm) |
|---|---|---|---|---|
| April   1 | 12 | 10 | — | $0 + 12 - 10 = 2$ |
| 8 | 12 | 8 | — | $2 + 12 - 8 = 6$ |
| 15 | 12 | 18 | — | $6 + 12 - 18 = 0$ |
| 22 | 12 | 9 | — | $0 + 12 - 9 = 3$ |
| 29 | 16 (approx) | 30 | — | $3 + 16 - 30 = 0$[4] |
| May   6 | 20 | 10 | — | $0 + 20 - 10 = 10$ |
| 13 | 20 | 20 | — | $10 + 20 - 20 = 10$ |
| 20 | 20 | 5 | — | $10 + 20 - 5 = 25$ |
| 27 | 20 | 10 | — | $25 + 20 - 10 = 35$ |
| June   3 | 25 | 30 | — | $35 + 25 - 30 = 30$ |
| 10 | 25 | 10 | — | $30 + 25 - 10 = 45$ |
| 17 | 25 | 20 | — | $45 + 25 - 20 = 50$[5] |
| 24 | 25 | 10 | 25 | $50 + 25 - 35 = 40$ |

1. Water loss by evapotranspiration can be based on the average potential transpiration rate for the month for the area (*see* Table 14.2).
2. Water gained by either rainfall (actual local raingauge figure) or irrigation.
3. Soil moisture deficit is the balance after the previous week's balance is adjusted for the net gain/loss of water. It is assumed that, unless there is information to the contrary, SMD in the UK at beginning April is zero.
4. SMD cannot be less than zero because water above field capacity drains away.
5. 25 mm water added as a 50 mm SMD is reached (irrigation would be postponed if a period of heavy rainfall was forecast).

ment but it is essential that the rate of application is related to soil infiltration rate and the droplets size in the spray should not be large enough to damage the surface structure (*see* tilth). Indoors, spray lines fitted above the plants can lead to very high humidity levels and wet foliage, predisposing some plants to disease (*see* grey mould) and should be restricted to watering low level crops, e.g. lettuce, deliberately increasing humidity, or winter flooding (*see* conductivity). **Trickle** lines deliver water very slowly to the soil, leaving plant foliage and the soil surface dry, which ensures a drier atmosphere and reduced water loss. However, care is needed because there is very little sideways spread of water into coarse sand, loose soil, or a growing medium that has completely dried out. **Drip** irrigation is a variation on the trickle method but the water is applied through pegged-down thin, flexible 'spaghetti' tubes to exactly where it is needed, e.g. in each pot or base of each plant.

**Capillary benches** can be used to keep pot plants watered. The pots stand on, and the contents make contact with, a level 50 mm bed of sand kept saturated at the base by an automatic water supply. Water lost by evaporation at the surface and from the plant is replaced by **capillary rise**. The sand must be fine enough to lift the water but coarse enough to ensure that the flow rate is sufficient. **Capillary matting** made of fibre woven to a thickness and pore size to hold, distribute, and/or lift water has many uses in watering containerized crops indoors or outside. Containers with built-in water reserves and easy watering systems utilize capillarity to keep the rooting zone moist, as shown in Figure 17.1.

## WATER QUALITY

Water used in horticulture is taken from different sources and has different dissolved impurities. Soft water has very few impurities whereas **hard water** contains large quantities of calcium or magnesium salts which raise the pH of the growing medium, especially where the impurities accumu-

late (*see* lime). Even small quantities of micro-elements such as boron or zinc have to be allowed for when making up nutrient solutions which are to be re-circulated (*see* hydroponics). Water taken from boreholes in coastal areas can have high concentrations of salt which can lead to **salt concentration problems**. Water drawn from rivers, lakes, or even on-site reservoirs may contain algal, bacterial or fungal pollution which can lead to blocked irrigation lines or plant disease (*see* hygienic growing).

## WATER CONSERVATION

The need to manage water efficiently is a major concern in the use of scarce resources. Responsible action is increasingly supported by legislation and the higher price of using water. Clearly the major factors which determine the level of water use are related to the choice of plant species to be grown and the reasons for growing. The selection of drought-tolerant rather than sensitive plantings is fundamental. Likewise the recycling of water and the capture of rainwater are important considerations in the choice of water source. Some growing systems are inherently less water intensive, but in most of them there are many ways in which water use can be reduced if certain principles are kept in mind and acted upon appropriately.

Water is lost primarily through drainage or evaporation. Where application is partially controlled, the correct relationship between water applied and water holding helps to prevent leaching, a process which also causes nutrient loss (*see* water holding capacity). Thus returning an outdoor soil to less than field capacity helps avoid losses to drainage and by runoff in the event of unexpected rainfall.

Evaporation losses can be minimized by reducing the action of the drying atmosphere. Overhead application of water in the open can be limited by increasing the growing medium's water reservoir. Soils have different available water holding capacities but most can be improved by the

addition of suitable organic matter. Most importantly, the rooting depth can be increased by maintaining good soil structure. Plants should be
encouraged to establish as quickly as possible but,
after the initial watering-in, infrequent applications will encourage the plant to put down deep
roots by searching for water. When water does
have to be applied overhead, this should be undertaken in cool periods. However, care should be
taken to avoid creating conditions which encourage diseases such as *Botrytis*.

Water is lost more rapidly from a moist than a
dry soil surface. After just 5 mm of water has
been lost from the surface, the rate of evaporation
falls significantly. Infrequent application thus
helps, but even more effective is the delivery of
water to specific spots next to plants (*see* seep-
hose) or from below through pipes to the rooting
zone. Avoid bringing moist soil to the surface. If
hoeing is undertaken it should be confined to the
very top layers; this also reduces the risk of root
damage. Losses from the surface can be reduced
considerably by plant cover and almost eliminated
by the use of mulches. Loss of water from the
plants themselves is reduced when they are
grouped together rather than spaced out.

Unless maximum growth rates are the main
consideration, reduced application saves water,
money and staff time without detriment to most
plantings. In production horticulture, the introduction of sophisticated moisture-sensing equipment and computer control has enabled water to
be delivered more precisely when and where it is
needed. This has led to considerable reductions
in water use.

## PRACTICAL EXERCISES

1   Construct a **raingauge** and keep a record of
rainfall in your locality. Compare the figures
with yearly and seasonal averages for your
locality and those of other areas.

2   Keep a **water balance sheet** using your own
rainfall figures and the potential transpiration
rates of your locality. Note when irrigation is
needed for a given situation (*see* Table 14.3).

3   Allow seeds to germinate in each of three
polythene bags two-thirds full of moist loamy
soil. When the seeds have germinated, punch
holes in two bags and thoroughly wet all
three soils regularly. When plants are well
grown put one of the bags with holes into an
unpunched polythene bag; continue to keep
soils well watered. Examine root systems at
the end of several weeks. Repeat with different
plants, including weeds.

4   Cut the bottom out of a clear plastic bottle and
punch holes in the cap. Secure the bottle
upside down and almost fill with dry soil sieved
through a 2 mm mesh. Add a known amount
of water and observe the **wetting of the dry
soil**. Catch any water that drains through in a
beaker. Note how long it takes to wet the soil.
Allow the fully wetted soil to drain completely,
then calculate the moisture-holding capacity
of the soil. Repeat with other soils and compost ingredients and note differences.

5   Fill small plant pots with different growing
media and plant sunflower seeds. Keep moist
until all have germinated then wet thoroughly
and allow to drain. Do not water again but
keep the plants in the same environment until
water supply is used up. Note when each wilts
and record when each reaches the permanent
wilting point. Note the difference in **available
water** content.

6   Take a ball of dry soil and moisten it until it
becomes soft and **friable**. Continue to add
water until it becomes **plastic** and finally **fluid**.
Repeat with different soil textures and note
how they differ. Try and keep a record of the
**working days** of different soils, i.e. days when
cultivation can be undertaken successfully.

7   Dig **profile pits** and assess the drainage qualities of the soil by looking at soil structure,
rooting, soil colours etc.

8   Examine local ditches after winter rains have
returned soils to field capacity and note **drain
outfalls**. After a period of rain, account for
any drains not running.

## FURTHER READING

Baily, R. *Irrigated Crops and Their Management* (Farming Press, 1990).

Castle, D.A., McCunnell, J., and Tring, I.M. *Field Drainage: Principles and Practice* (Batsford Academic, 1984).

Farr, E., and Henderson, W.C. *Land Drainage* (Longman Handbooks in Agriculture, 1986).

Hope, F. *Turf Culture: A Manual for the Groundsman* (Cassell, 1990).

MAFF Advisory Booklets:

No. 2067 *Irrigation Guide* (1979).

No. 2140 *Watering Equipment for Glasshouse Crops* (1980.

MAFF Advisory Leaflets:

Nos 721–740 *Getting Down to Drainage* (various dates).

No. 776 *Water Quality for Crop Irrigation* (1981).

MAFF Advisory Reference Book No. 138 *Irrigation* (1976).

MAFF Technical Bulletin No. 34 *Climate and Drainage* (1970).

# 15

# Soil Organic Matter

*Organic matter exerts a profound influence on crop nutrition, soil structure, and cultivations. In this chapter the three main categories of organic matter are described; the living organisms, dead but identifiable organic matter, and humus. The nutrient cycles describe how minerals are released from plant tissue and made available in the soil as plant nutrients. The interlocked nature of the different nutrient cycles is illustrated by considering the effect of the carbon : nitrogen ratio. The organic matter levels in different soils and the factors that cause the variations are examined. The beneficial effects of organic matter are established and the characteristics of different sources of bulky organic matter are considered.*

## LIVING ORGANISMS IN THE SOIL

As in any other plant and animal community the organisms that live in the soil form part of the **food webs**. The main types present in any soil are the **primary producers** which are those capable of utilizing the sun's energy directly, synthesizing their own food by photosynthesis, such as green plants (*see* photosynthesis), the **primary consumers** which are those organisms which feed directly on plant material, and **secondary consumers** which feed only on animal material. In practice, there are some organisms which feed on both plants and animals and also parasites living on organisms in all categories.

**Decomposers are an important group which have the special function within a community of breaking down dead or decaying matter into simpler substances with the release of inorganic salts, making them available once more to the primary producers. Primary decomposers** are those organisms which attack the freshly dead organic matter. These include insects, earthworms and fungi. Fungi are particularly important in the initial decomposition of fibrous and woody material. **Secondary decomposers** are those organisms which live on the waste products of other decomposers and include bacteria and many species of fungi.

### Plant roots

These are important as contributors to the organic matter levels in the soil. They shift and move soil particles as they penetrate the soil and grow in size. This rearrangement changes the sizes and shapes of soil aggregates and when these roots die and decompose, a channel is left which provides drainage and aeration. Root channels are formed over and over again unless the soil becomes too dense for roots to penetrate. Roots absorb water from soils and dry it, causing those with a high clay content to shrink and crack. This helps develop and improve structures on heavier soils (*see* roots).

### Earthworms

There are ten common species of earthworm in Britain which vary in size from *Lumbricus terrestris*, which can be in excess of 25 cm, to the many small species less than 3 cm long. The main food of earthworms is dead plant remains. Casting species of earthworms are those which eat soil as well as organic matter and their excreta consists of intimately mixed, partially digested, finely divided organic matter and soil. Many species never produce casts and only two species regularly cast on the surface giving the **worm casts** which are a problem on fine grass areas, particularly in the autumn. It has been estimated that in English pastures the production of casts each year is 20–40 t/ha, the equivalent of 5 mm of soil deposited annually. This surface casting also leads to the incorporation of the leaf litter and the burying of stones. However, *L. terrestris* is the organism mainly responsible for the **burying** of large quantities of litter by dragging plant material down its burrows. Earthworms make burrows by pushing between cracks and some by eating soil. The **network of burrows** which develops as a result of worm activity is an important factor in maintaining a good structure, particularly in uncultivated areas and in soils of a low clay content. Some species live entirely in the surface layers of the soils, others move vertically establishing almost permanent burrows down to two metres.

**The distribution and activity of earthworms are largely governed by moisture, soil pH, temperature, organic matter, and soil type.** Most species tend to be more abundant in soils where there are good reserves of calcium. Earthworm populations are usually lower on the more acid soils and most species do not survive on soils less than pH 4.5. Worm numbers decrease in dry conditions, but they can take avoiding action by burrowing to more moist soil or by hibernating. Each species has its optimum temperature range. For *L. terrestris* this is about 10°C which is typical of soil temperatures in the spring and autumn in the UK. Soils with low organic matter levels support only small populations of worms. In contrast compost heaps and stacks of farmyard manure have high populations. In oak and beech woods where the fallen leaves are palatable to worms their populations are large and they can remove a high proportion of the annual leaf-fall. This also happens in orchards unless earthworm populations have been reduced by harmful chemicals such as copper. Light and medium loams support a higher total population than clays, peats, and gravelly soils.

**Slugs** and **snails**, **arthropods** such as millipedes, springtails, and mites and **nematodes** are also found in high numbers and play an important part in the decomposition of organic matter. Several species are also horticultural pests (*see* Chapter 10).

### Bacteria

**Bacteria are present in soils in vast numbers. About 1000 million or more occur in each gram of fertile soil. Consequently, despite their microscopic size, the top 150 mm of fertile topsoil carries about one tonne of bacteria per hectare. There are many different species of bacteria to be found in the soil and most play a part in the decomposition of organic matter.** Many bacteria attack minerals; this leads to the weathering of rock debris and the release of plant nutrients. **Detoxification** of pesticides and herbicides is an important activity of the bacterial population of cultivated soils.

Soil bacteria are inactive at temperatures below 6°C but their activities increase with rising temperatures up to a maximum of 35°. Actively growing bacteria are killed at temperatures above 82°C, but several species can form thick-walled resting spores under adverse conditions. These spores are very resistant to heat and they survive temperatures up to 120°C. **Partial sterilization** of soil can kill the actively growing bacteria but not the bacterial spores. The growth rate and multiplication depend also upon the **food supply**. High organic matter levels support high bacteria populations so long as a balanced range of nutrients is present. Bacteria thrive in a range of **pH 5.5–7.5**; fungi tend to dominate the more acid soils. Aerobic conditions should be maintained because the beneficial organisms as well as plant roots require oxygen, whereas many of the bacteria which thrive under anaerobic conditions are detrimental.

### Fungi

The majority of fungi live saprophytically on soil organic matter. Some species are capable only of utilizing simple and easily decomposable organic matter whereas others attack cellulose as well. Among fungi are an important group which decompose lignin, making them one of the few primary decomposers of wood and fibrous plant material. Several fungi in the soil are parasites and examples of these are discussed in Chapter 11. Fungi appear to be able to tolerate acid conditions and low calcium better than other micro-organisms and are abundant in both neutral and acid soils. Most are well adapted to survive in dry soils but few thrive in very wet conditions. Their numbers are high in soils rich in plant residues but decline rapidly as the readily decomposable material disappears. The bacteria persist longer, where present, and eventually consume the fungal remains.

### The rhizosphere

The rhizosphere is a zone in the soil with no real boundary which is influenced by roots. Living roots change the atmosphere around them by using up oxygen and producing carbon dioxide (*see* respiration). Roots exude a variety of organic compounds which hold water and form a coating that bridges the gap between root and nearby soil particles. Micro-organisms occur in greatly increased numbers and are more active when in proximity to roots. Some actually invade the root cells where they live as **symbionts** (*see* Chapter 11). The *Rhizobium spp.* of bacteria live symbiotically with many legumes. **Symbiotic associations involving plant roots and fungi are known as mycorrhiza.**

There is considerable interest in exploiting the potential of **mycorrhiza** which appear to be associated with a high proportion of plants. The fungus obtains its carbohydrate requirements from the plant and the plant gains greater access to nutrients in the soil, especially phosphates, through the increased surface area for absorption and because the fungus appears to utilize sources not available to higher plants. Most woodland trees have fungi covering their roots and penetrating the epidermis. Orchids and heathers have an even closer association in which the fungi invade the root and coil up within the cells. The association appears to be necessary for the successful development of the seedlings. Mycorrhizal plants generally appear to be more tolerant of transplanting and this is thought to be an important factor for orchard and container grown ornamentals.

### NUTRIENT CYCLES

All the plant nutrients are in continuous circulation between plants, animals, the soil and the air. The processes contributing to the production of simpler inorganic substances such as ammonia, nitrites, nitrates, sulphates and phosphates are sometimes referred to as **mineralization**. Mineralization yields chemicals which are readily taken up by plants from the soil solution. The formation of humus, organic residues of a resistant nature, is known as **humification**. Both mineralization and humification are intimately tied up in the same decomposition process but the terms help identify the end product being studied. Likewise it is possible to follow the circulation of carbon in the carbon cycle and nitrogen in the nitrogen cycle although these nutrient cycles along with all the others are interrelated.

## Carbon cycle

**Green plants obtain their carbon from the carbon dioxide in the atmosphere and during the process of photosynthesis are able to fix the carbon, converting it into sugar. Some carbon is returned to the atmosphere by the green plants themselves during respiration, but most is incorporated into plant tissue. The carbon incorporated into the plant structure is eventually released as carbon dioxide, as illustrated in Figure 15.1.**

All living organisms in this food web release carbon dioxide as they respire. The sugars, cellulose, starch and proteins of **succulent** plant tissue, as found in young plants, are rapidly decomposed to yield plant nutrients and have only a short-term effect. In contrast, the **lignified tissue** of older plants rots more slowly. Besides the release of nutrients, **humus** is formed from this fibrous and woody material, which has a long-term effect on the soil. Plants grown in the vicinity of vigorously decomposing vegetation, e.g. cucumbers in straw bales, live in a carbon dioxide enriched atmosphere. Carbon dioxide is also released on combustion of all organic matter, including the fossil fuels such as coal and oil. Organic materials such as paraffin or propane which do not produce harmful gases when burned cleanly are used in protected culture for carbon dioxide enrichment.

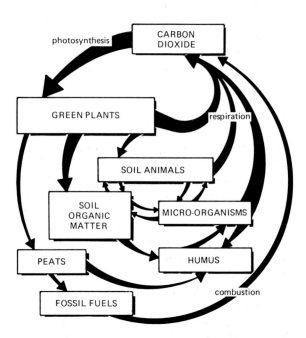

**Figure 15.1** *Carbon cycle*
*The recycling of the element carbon by organisms is illustrated. Note how all the carbon in organic matter is eventually released as carbon dioxide by respiration or combustion. Green plants convert the carbon dioxide by photosynthesis into sugars which form the basis of all the organic substances required by plants, animals and micro-organisms.*

## Nitrogen cycle

**Plants require nitrogen to form proteins. Although plants live in an atmosphere largely made up of nitrogen they cannot utilize gaseous nitrogen. Plants take up nitrogen in the form of nitrates and, to a lesser extent, as ammonia. Both are released from proteins by a chain of bacterial reactions as shown in Figure 15.2.**

**Ammonifying bacteria** convert the proteins they attack to ammonia. Ammonia from the breakdown of protein in organic matter or from **inorganic nitrogen** fertilizers is converted to nitrates by **nitrifying bacteria**. This is accomplished in two stages. Ammonia is first converted to **nitrites** by *Nitrosomonas spp.* Nitrites are toxic to plants in small quantities but they are normally converted to nitrates by *Nitrobacter spp.* before they reach harmful levels. Ammonifying and nitrifying bacteria thrive in aerobic conditions but where there is no oxygen, anaerobic organisms dominate. Many anaerobic bacteria utilize nitrates and in doing so convert them to **gaseous nitrogen**. This **denitrification** represents an important loss of nitrate from the soil which is at its most serious in well-fertilized, warm and waterlogged land.

Although gaseous nitrogen cannot be utilized by plants it can be converted to plant nutrients by some micro-organisms. *Azotobacter* are free-living bacteria which obtain their nitrogen requirements from the air. As they die and decompose, the nitrogen trapped as protein is converted to

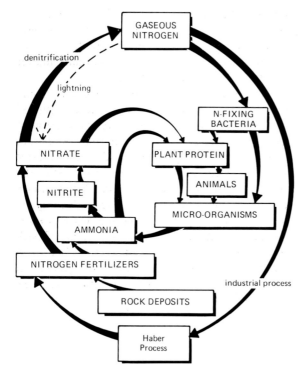

**Figure 15.2** *Nitrogen cycle.*
*The recycling of the element nitrogen by organisms is illustrated. Note the importance of nitrates which can be taken up and used by plants to manufacture protein. Micro-organisms also have this ability but animals require nitrogen supplies in protein form. Gaseous nitrogen only becomes available to organisms after being captured by nitrogen-fixing organisms or via nitrogen fertilizers manufactured by man. In aerobic conditions ammonia in the soil is converted by soil bacteria to nitrates (nitrification) whereas in anaerobic conditions nitrates are reduced to nitrogen gases.*

ammonia and then nitrates by other soil bacteria. The *Rhizobia* which live in root nodules on some legumes also trap nitrogen to the benefit of the host plant. Finally, nitrogen gas can be converted to ammonia industrially in the **Haber process** which is the basis of the artificial nitrogen fertilizer industry.

**Sulphur cycle**

Sulphur is an essential constituent of plants which accumulates in the soil in organic forms. This sulphur does not become available to plants until the organic form is mineralized by aerobic micro-organisms which yield soluble **sulphates**. Under anaerobic conditions there are micro-organisms which utilize organic sulphur and produce hydrogen sulphide.

**C : N ratio**

All nutrients play a part in all nutrient cycles simply because all organisms need the same range of nutrients to be active. Normally there are adequate quantities of nutrients with the exception of carbon or nitrogen which are needed in relatively large quantities. **A shortage of nitrogenous material would lead to a hold-up in the nitrogen cycle but would also slow down the carbon cycle, i.e. the decomposition of organic matter is slowed because the micro-organisms concerned suffer a shortage of *one* of their essential nutrients. A useful way of expressing the relative amounts of the two important plant foods is in the carbon : nitrogen ratio.** Plant material has relatively wide C : N ratios but those of micro-organisms are much narrower. Three quarters of the carbon in plants is utilized by micro-organisms during decomposition as an energy source and released as carbon dioxide, whereas all the nitrogen used is incorporated in the microbial body protein.

Sometimes the C : N ratio is so wide that some nitrogen is drawn from the soil and 'locked up' in the microbial tissue. This is what happens when **straw** which has a 60 : 1 ratio is dug into a soil. For example if one thousand 12 kg bales of straw are dug into one hectare of land then the addition to the soil will be 12 000 kg of straw containing 4800 kg of carbon and 80 kg of nitrogen. Three-quarters of the carbon (3600 kg) is utilized for energy and lost as carbon dioxide and a quarter (1200 kg) in incorporated over several months into microbial tissue. Microbial tissue has a C : N ratio of about 8 : 1 which means that by the time the straw is used up some 150 kg of nitrogen is locked up with the 1200 kg of carbon in the micro-organisms. Since there was only 80 kg of nitrogen in the straw put on the land, the other 70 kg. has been 'robbed' from the soil. This nitrogen is rendered unavailable to plants ('locked up') until the micro-organisms die and decompose. To ensure rapid decomposition or to prevent a detrimental effect on crops the addition of straw must be accompanied by the addition of nitrogen. Nitrogen is released during decomposition if the organic material has a C:N ratio narrower than 30:1, such as young plant material, or with nitrogen-supplemented plant material such as farmyard manure (FYM). In general, fresh organic matter decomposes very rapidly so long as conditions are right, but the older residues tend to decompose very slowly.

## HUMUS

The dead organic matter of the soil is gradually decomposed by active micro-organisms until it consists of finely divided plant and micro-organism remains. **This process of humification leads to the formation of humus which is the slowly decomposing residues of soil organic matter. It is a black colloidal material which coats soil particles and gives topsoil its characteristic dark colour.** It is mainly derived from the decomposition of fibrous vegetation such as straw which is rich in **lignin**. This colloidal material has a high **cation exchange capacity** and therefore can make a major contribution to the retention of exchangeable cations, especially on soils low in clay (*see* sands). It also adheres strongly to mineral particles, especially clay, which makes it a valuable agent in soil **aggregation** and soil **workability**. Humus is eventually decomposed by bacteria and so the amount in the soil is very dependent on the addition of appropriate bulky organic matter.

## ORGANIC MATTER LEVELS

The routine laboratory method for **estimating organic matter** levels depends upon finding the total carbon content of the soil. A simpler method which can be used is to dry a sample of soil and

burn off the organic matter. After cooling, the soil can be re-weighed and the loss in weight represents destroyed organic matter. These methods give an overall total of soil organic matter excluding the larger soil animals.

Most topsoil contains between 1 per cent and 6 per cent of organic matter, whereas subsoil contains less than 2 per cent. The distribution of organic matter under grass in normal temperate areas is shown in Figure 15.3. Soil organic matter is concentrated in the topsoil because most of the roots occur in this zone and the plant residues tend to be buried in the upper layers. **The organic matter level in any part of the soil depends upon how much fresh material is added compared with the rate of decomposition. It is stable when these two processes are balanced and the equilibrium reached is determined mainly by climate, soil type and treatment under cultivation.**

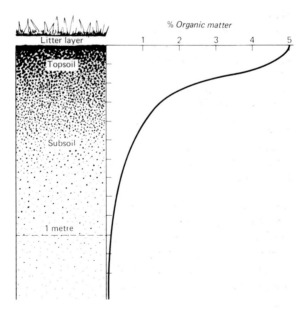

**Figure 15.3** *Distribution of organic matter in soil. Organic matter content of soil decreases from the soil surface downwards. Note that the topsoil is significantly richer in humus, which gives it a characteristically darker colour.*

**Climate**

Climate affects both the amount of organic matter added and the rate of decomposition. Below 6°C there is no microbial activity, but it increases with increasing temperature so long as conditions are otherwise favourable. In dry areas there is not only less crop growth, resulting in less organic matter being added to the soil, but also less microbial action. In **warm climates**, where there is adequate moisture, low organic matter results from very much increased decomposition. In **cooler areas** there tends to be an accumulation of organic matter because the decreased plant growth is more than offset by the reduced micro-organism activity which occurs over the long winter periods. Organic matter also tends to accumulate in **wetter conditions**. Where waterlogging is prevalent, 10−20 per cent organic matter levels develop; where waterlogging is permanent, organic matter accumulates to give rise to **peat**.

**Soil type**

Generally, soils with higher clay contents have higher organic matter levels. Coarse sands and sandy loams tend to be warmer than finer textured soils and have better aeration, which results in higher microbial activity. Such soils often support less plant growth because of poor fertility and poor water holding capacity. These factors combine to give soils with low organic matter levels. In cultivation these same soils become a problem unless large quantities of organic matter are applied at frequent intervals to maintain adequate humus levels. Such soils are often referred to as 'hungry soils' because of their high demand for manure. On finer-textured soils the higher fine sand, silt or clay content increases water holding capacity. This reduces soil temperature, resulting in less microbial activity. The presence of clay directly reduces the rate of decomposition because it combines with humus and protects it from microbial attack.

**Cultivation**

On first cultivation the increased aeration and nutrients stimulate micro-organisms and a new equilibrium with lower soil organic matter levels prevails. Once under cultivation, grasses and high-producing legumes tend to increase levels but most crops, particularly those in which complete plant removal occurs, lead to decreased organic matter levels. Only large, regular dressings of bulky organic matter can improve or maintain the level of soil organic matter on cultivated soils.

Organic matter can accumulate under grass and form a mat on the surface where the carbon cycle is slowed because of nutrient deficiency such as phosphates, or surface soil acidity.

**Organic soils**

While all soils contain some organic matter, most are classified as mineral soils. Organic soils are those that have enough organic matter present to dominate the soil properties and they develop where decomposition is slow because the activity of micro-organisms is reduced by cold or water-logged conditions. **Peat** is formed of partially decomposed plant material. This occurs in water-logged conditions, usually low in nutrients, where decomposition rates are low. There are great differences between peats because of the variations in conditions where they occur and the species of plants from which they are formed. Some peat is formed in shallow water, as found in poorly drained depressions or infilling lakes. In such circumstances the water drains from surrounding mineral soils and consequently has sufficient nutrients to support vegetation often dominated by sedges, giving rise to **sedge peat**. **Sphagnum moss peat** is usually formed in cold conditions where vegetation is kept continuously wet because of high rainfall and a moisture retentive surface. The water present has a very low mineral content. The dead vegetation becomes very acid and decomposes slowly.

Some of the peatlands which are enriched with

minerals prove very valuable when drained. Many of these soils are used to produce vegetables and other high value crops. Unfortunately the increased aeration allows the organic matter to be decomposed at a rate faster than it can be replenished and when the surface dries out the light particles are vulnerable to wind erosion. Consequently the soil level of these areas is falling at the rate of three metres every hundred years. Subsidence can be minimized by keeping the water table as high as possible and providing protection against wind erosion.

## BENEFITS OF ORGANIC MATTER

The presence of partially decomposed organic matter creates an **open soil structure** and, on many soils, increases water holding capacity. As it decomposes it acts as a **slow release fertilizer**. The **humus** coats soil particles and modifies their characteristics. On sandy and silty soils the humus enables stable crumbs to be formed. Stable, well-structured fine sands and silts are only possible under intensive cultivation if high humus levels are maintained by the addition of large quantities of bulky organic matter. The surface charges on humus are capable of combining with the clay particles, thereby making heavy soils less sticky and more friable. These surface charges also enable humus to hold cations against leaching: very important in soils low in clay. Humus also improves soil **water holding properties**. Its darkening effect increases heat absorption.

The **living organisms** in the soil play their part in the conversion of plant and animal debris to minerals and humus. The *Rhizobia* and *Azotobacter spp.* fix gaseous nitrogen while other bacteria play an important role in the detoxification of harmful organic materials such as pesticides and herbicides. Soil structure is modified by the influence of plant roots and earthworms.

## ADDITION OF ORGANIC MATTER

It is normal in horticulture to return **residues** to cultivated areas where possible. Whether or not the plant remains are worked into the soil in which they have been grown depends upon their nature. The residues of some crops such as tomatoes in the greenhouse are removed to reduce disease carry-over and because they cannot easily be worked into the soil. Other crops such as hops are removed for harvesting and some of the processed remains, spent hops, can be returned or used elsewhere. Wherever organic matter is removed, whether it is just the marketed part such as top fruit from the orchard or virtually the whole crop such as cucumbers from a greenhouse, the nutrients removed must be replaced to maintain fertility (*see* fertilizers). Although some forms do not return nutrients to the soil, **bulky organic matter** such as compost, straw, farmyard manure and peat is an important means of maintaining organic matter and humus levels. The main problem is obtaining cheap enough sources because their bulk makes transport and handling a major part of the cost. They can be evaluated on the basis of their nutrient value and their effect on the physical properties of soil as appropriate.

### Composting

Composting refers to the rotting down of plant residues before they are applied to soils. Many gardeners depend on composting as a means of using garden refuse to maintain organic matter levels in their soils. On a larger scale there is interest in the use of composted **town refuse** for horticultural purposes. Horticulturists are increasingly concerned with the recycling of wastes and attention is being given to modern composting methods. For successful composting, conditions must be favourable for the decomposing micro-organisms. The material must be moist and well-aerated throughout. As the heap is built, separate layers of lime and nitrogen are added as necessary to ensure the correct pH and C : N ratio. Organic waste brought together in large enough quantities under ideal conditions can generate enough heat to take the temperature to over 70°C within seven days. If all the material is brought to this temperature it has the advantage of killing harmful organisms and weed seeds.

### Straw

Straw is an agricultural crop residue readily available in most parts of the country, but care should be taken to avoid straw with harmful **herbicide residues**. It is ploughed in or composted and then worked in. There appears to be no advantage in composting if allowance is made for the demand on nitrogen by soil bacteria. About 6 kg of nitrogen fertilizer needs to be added for each tonne of straw for composting or preventing soil robbing. Chopping the straw facilitates its incorporation and while undecomposed it can open up soils. On decomposition it yields very little nutrient for plant use but makes an important contribution to maintaining soil humus levels. Straw bales suitably composted on site are the basis of producing an open growing medium for cucumbers.

### Farmyard manure (FYM)

This is the traditional material used to maintain and improve soil fertility. It consists of straw, or other bedding, mixed with animal faeces and urine. The exact value of this material depends upon the proportions of the ingredients, the degree of decomposition and the method of storage. Samples vary considerably. Much of the FYM is rotted down in the first growing season but almost half survives for another year and half of that goes on to a third season and so on. A full range of nutrients is released into the soil and the addition of the major nutrients should be allowed for when calculating **fertilizer requirements**. The continued release of large quantities of nitrogen can be a problem, especially on unplanted ground in the autumn, when the nitrates formed are leached deep into the soil over the winter.

FYM is most valued for its ability to provide the organic matter and humus for maintaining or improving soil structure. As with any other bulky organic matter, FYM must be worked into soils where conditions are favourable for its continued decomposition. Where fresh organic matter is worked into wet and compacted soils the need for oxygen outstrips supply and anaerobic conditions develop to the detriment of any plants present. Where this occurs a foul smell (*see* sulphur) and grey colourings occur. FYM should not be worked in deep, especially on heavy soil.

### Horticultural peats

**Sphagnum moss peats** have a fibrous texture, high porosity, high water retention and a low pH. They are used extensively in horticulture as a source of bulky organic matter and are particularly valued as an ingredient of potting composts because with their stability, excellent porosity and high water retention they can be used to create an almost ideal root environment.

**Sedge peats** tend to contain more plant nutrients than sphagnum moss. They are darker, more decomposed, and have a higher pH level but a slightly lower water holding capacity. They tend to be used for making peat blocks.

Considerable efforts are being made to find alternative materials to replace peat in order to avoid destroying valued wetland habitats.

### Leaf mould

Leaf mould is made of rotted leaves of deciduous trees. It is low in nutrients because nitrogen and phosphate are withdrawn from the leaves before they fall and potassium is readily leached from the ageing leaf. They are often composted separately and much valued in ornamental horticulture for a variety of uses. However, unless they are from trees growing in very acid conditions the leaves are rich in calcium and should not be used with calcifuge plants.

**Pine needles** are covered with a protective layer which slows down decomposition. They are low in calcium and the resins present are converted to acids. This extremely acid litter is almost resistant to decomposition. It is valued in the propagation and growing of calcifuge plants such as rhododendrons and heathers and as a material for constructing decorative pathways.

### Air-dried digested sludge

This consists of sewage sludge which has been fermented in sealed tanks, drained, and stacked to dry. The harmful organisms and the objectionable smells of raw sewage are eliminated in this process. It provides a useful source of organic matter but is low in potash. Advice should be taken before using sewage sludges because is some regions they contain large quantities of metals such as zinc, nickel and cadmium, which can accumulate in the soil to levels toxic to plants.

### Leys

The practice of ley farming involves grassing down areas and is common where arable crop production can be closely integrated with livestock. At the end of the ley period the grass or grass and clover sward is ploughed in. The root action of the grasses and the increased organic matter levels can improve the structure and workability of problem soils. There are some pest problems peculiar to cropping after grass which should be borne in mind (*see* wireworms), and generally the ley enterprise has to be profitable in its own right to justify its place in a horticultural rotation. It is practised in some vegetable production and nursery stock areas.

### Mulching

**Mulches are materials applied to the surface of the soil to suppress weeds, modify soil temperatures, reduce water loss, protect the soil surface and reduce erosion.** Many organic materials are used for this purpose, including straw, leaf mould, peat, compost, lawn mowings and spent mushroom compost. The organic mulches increase earthworm activity at the surface, which promotes better and more stable soil structure in the top layers. Soil compaction by water droplets is reduced and, as the organic mulches are incorporated, the soil structure is improved. If thick enough, mulches can suppress weed growth but it is counterproductive to introduce a material that contains weeds. Likewise, care should be taken not to introduce pests and disease or use a material such as compost where slugs can be a problem. In effect, the mulch is acting as an extra layer of loose soil. Thus, water loss from the soil surface is reduced because it is covered with a dry layer (*see* evaporation). Soil temperatures lag behind the surface temperatures with the greater lag at greater depth. Depending on their insulatory properties, mulches can reduce soil temperatures in the summer but retain warmth later in the autumn. Manufactured materials such as paper, metal foil or, most commonly, polythene are used. The colour of the mulch is important because light coloured material will reflect radiation whereas dark material can lead to earlier cropping by warming up the soil earlier. These materials are particularly effective in reducing water loss by evaporation at the surface (*see* water conservation).

## ORGANIC PRODUCTION

Organic production in the European Community is subject to regulations laid down in 1991. These require that the fertility and the biological activity of the soil must be maintained or increased by the cultivation of legumes, green manures or deep-rooting crops in an appropriate multiannual rotation and by the incorporation of organic matter including farmyard manure, composted or not, from holdings producing according to the same regulations. If adequate nutrition or soil conditioning cannot be achieved by these means then other, specified, organic or mineral fertilizers, as listed in Table 15.1, may be applied. For compost activation, appropriate micro-organisms or plant-based preparations may be used. These organic or mineral fertilizers may be applied only to the extent that the adequate nutrition of the crop being rotated, or soil conditioning, is not possible by the preferred methods.

**Table 15.1**　Sources of nutrients for use in organic growing

| |
| --- |
| Farmyard and poultry manure |
| Slurry or urine |
| Composts from spent mushroom and vermiculture substrates |
| Composts from organic household refuse |
| Composts from plant residues |
| Processed animal products from slaughterhouses and fish industries |
| Organic byproducts of foodstuffs and textile industries |
| Seaweeds and seaweed products |
| Sawdust, bark and wood waste |
| Wood ash |
| Natural phosphate rock |
| Calcinated aluminium phosphate rock |
| Basic slag |
| Rock potash |
| Sulphate of potash* |
| Limestone |
| Chalk |
| Magnesium rock |
| Calcareous magnesium rock |
| Epsom salt (magnesium sulphate) |
| Gypsum (Calcium sulphate) |
| Trace elements (boron, copper, iron, manganese, molybdenum, zinc)* |
| Sulphur* |
| Stone meal |
| Clay (bentonite, perlite) |

\* Need recognized by control body.

## PRACTICAL EXERCISES

1　**Earthworms**, particularly *Lumbricus terrestris*, can be brought to the surface by a 0.5% solution of formalin (use 25 ml of 40% formalin in 5 litres of water). Apply 10 litres from a watering can over a marked-out one metre square. Count the earthworms that appear, wash in clean water and release on clean land. Compare the numbers of different types of soil, cultivated areas, woodland, etc. (Do **not** use this method on decorative grass; the grass turns yellow but does recover.)

2　Take a handful of chopped **straw** and mix with a neutral fertile soil in a polythene bag with drainage holes punched in the bottom. Keep moist and warm. Note how long it takes for the straw to decompose. Repeat with different types of soil, including washed sand, acid soils and alkaline soils. Make up one bag with no drainage holes and keep soil saturated.

3　Examine the roots of legumes and note the occurrence of root nodules caused by *Rhizobia*.

4　Dry some soil in an oven. Take 100 grammes of dried soil and heat to red heat in a tin lid over a gas flame for an hour. Allow to cool, then re-weigh. The loss in weight roughly represents the **percentage of organic matter** in the soil. Repeat for different soils and at different depths.

5　Take a few soil crumbs shaken from grass roots and place in a saucer. Gently run in water until soil is half covered. Observe the structures after water has soaked into the soil, after ten swirls of the vessel and after one hundred swirls. Repeat with soils of different textures and organic matter levels. Compare the **stability** (*see* page 137) of the soils and make special note of those in which slaking occurs.

6　Find out the local sources of **bulky organic matter** and its cost delivered and spread. Calculate the effect of 10 tonnes of the material, allowing for water content, on the organic matter level of one hectare of topsoil (one hectare of dry topsoil weighs 2000 tonnes).

7   Compost different materials in different mixtures and containers. Compare the products and test each for viable weed seeds.

## FURTHER READING

Advisory Leaflet 257 *Green Manuring* (MAFF, 1973).

Advisory Reference Book 210 *Organic Manures* (MAFF, 1976).

Bragg, N. *Peatland: Its Alternatives* (Horticultural Development Council, 1991).

Edwards, C.A. and Lofty, J.R. *Biology of Earthworms*, 2nd edition (Halsted Press, New York: 1977).

Gotaas, H.B. *Composting* (WHO Monograph Series 31, 1956).

Hills, L.D. *Organic Gardening* (Penguin Books Ltd., Harmondsworth, Middlesex: 1977).

Jackson, R.M. and Raw, F. *Life in the Soil* (Edward Arnold, London: 1966).

Robinson, D.W. and Lamb, J.G.D. (Ed). *Peat in Horticulture* (HEA/Academic Press, London: 1975).

Russell, E.J. *The World of the Soil* (Collins New Naturalist, London: 1957).

# 16

# Plant Nutrition

*Green plants require sixteen elements in order to grow normally. The essential minerals enter the plant tissue in the form of ions from the growing media or to a lesser extent through the leaves. In established plant communities such as forest or grassland the minerals are recycled through the complex food webs and the community develops according to the many factors which affect plant growth, including the net gain or loss of minerals. Removing a plant or part of it breaks the natural cycle and prevents minerals contained in the plant from returning to the growing medium for re-use. These minerals can be replaced in many different ways. Where only a small quality of nutrient is needed it can be supplied through suitable types of bulky organic matter which release sufficient nutrient when mineralized. It is often impracticable to supply large quantities of nutrients by the addition of vast quantities of bulky organic matter; organic or inorganic fertilizers, which are more concentrated sources of major plant nutrients, resolve this dilemma. Many growing media, especially those used in composts, are very deficient in minerals and therefore supplementation with plant nutrients is essential. Ensuring mineral uptake is a major concern. The roots must be able to find the nutrients (see soil structure) and soil pH and mineral balance should be adjusted to provide optimum availability of nutrients.*

*This chapter describes the nature of soil pH, its effect on nutrient uptake and the methods of changing it to maintain optimum growing conditions. The methods of applying nutrients are examined and the characteristics of major and minor ('trace') elements in growing media are outlined. The chapter is concluded by considering how to apply the correct quantities of nutrient to different plants.*

## CONTROL OF SOIL pH

Life in the soil is greatly influenced by the soil pH, either directly or indirectly. The pH scale is a means of expressing the degree of acidity or alkalinity. pH values less than 7 are acid and the lower the figure, the greater the acidity. Values greater than 7 indicate alkalinity and pH 7 is neutral. Temperate area soils usually lie between pH 4 and 8: the vast majority are between 5.5 and 7.5.

For ideal growing conditions most plants require a soil of about pH 6.5, which is slightly acid. At this point most of the plant nutrients are available for uptake by the roots. Alkaline conditions are usually created by the presence of large quantities of calcium ('lime') which interferes with the uptake and utilization of several of the plant nutrients. **Calcicoles**, or 'lime-loving' plants, have evolved a different metabolism and are tolerant of high soil pH. In contrast, the **calcifuge** or 'lime-hating' plants, such as Rhododendrons and some heathers, do not even tolerate the level of calcium in soils at pH 5.5. Consequently, they must be grown in more acid conditions. At very low pH levels some substances such as aluminium become available at levels that are toxic to plants.

### Acidity and alkalinity

Pure water is neutral, i.e. neither acid nor alkaline. It is made up of two hydrogen atoms and one oxygen atom, expressed as the familiar formula $H_2O$. Most of the water is made up of water molecules in which the atoms stay together, but a tiny proportion disassociate, i.e. they form ions. Equal numbers of positive ions, called cations, and negative ions, called anions, are formed. In the case of water, an equal number of hydrogen cations, $H^+$, and hydroxyl anions, $OH^-$, exist within the clusters of water molecules ($H_2O$). When, as in water, the concentration of hydrogen ions is the same as that of hydroxyl ions, there is neutrality. As the concentration of hydrogen ions is increased, the concentration of hydroxyl ions decreases and acidity increases. Likewise, addition of hydroxyl ions and a decrease in hydrogen ions leads to increased alkalinity.

Many compounds form ions as they dissolve and mix intimately with water to produce a solution. All acids release hydrogen ions when dissolved in water. Whereas a strong acid (such as hydrochloric acid) dissociates when dissolved, only a part of a weak acid (such as carbonic or acetic acid) breaks up into ions. Bases, such as caustic soda, dissolve in water and increase the concentration of hydroxyl ions.

The pH scale expresses the amount of acidity or alkalinity in terms of hydrogen ion concentration. In order to present the scale simply, negative logarithms are used: 'pH' is the negative loga-

rithm, the mathematical symbol for which is 'p', of the hydrogen ion concentration, abbreviated as 'H'. It is important to note that, as the scale is logarithmic, a one-unit change represents a tenfold increase or decrease in hydrogen ion concentration, and two units a hundred-fold change. Thus a solution of pH 3 is ten times more acid than one of pH 4, one hundred times more acid than pH 5, and a thousand times more than pH 6.

The **buffering capacity** of a substance is its ability to resist change in pH. Pure water has no buffering capacity: the addition of minute quantities of acid or alkali has an immediate affect on its pH. In contrast, the cation exchange capacity of clays reduces the effect because the hydrogen ions exchange with calcium ions on the clay's colloid surface. Since the number of hydrogen ions being released or absorbed is small compared with the clay's reserve, the pH changes very little. High-humus soils similarly have the advantage of a high buffering capacity. A related buffer effect is seen when acids, such as the carbonic acid of rain, are incorporated into soils with 'free' lime present: the acid dissolves some of the carbonate with no accompanying change in pH.

### Soil acidity

Soil acidity is mainly determined by the balance of hydrogen ions and basic ions. A clay particle with abundant hydrogen ions acts as a weak acid, whereas if fully charged with bases (such calcium, Ca) it has a neutral or alkaline reaction (*see* percentage base saturation). Consequently, soil pH is usually regulated by the presence of calcium cations; soils become more acid as calcium is leached from the soil faster than it is replaced by mineral weathering. This is the tendency in temperate areas where rainfall, which is a weak acid (*see* carbonic acid), exceeds evaporation over the year. Thus hydrogen and aluminium ions take over the soil's cation exchange sites and the pH falls. Soils with large reserves of calcium, such as those derived from chalky boulder clay, do not become acid because they are kept base-saturated. In contrast, calcium ions are readily leached from free-draining sands in high rainfall areas and these soils tend to go acid rapidly.

In addition to the carbonic acid in rainfall, there are several other sources of acid that affect the soil. **Acid rain** (polluted rain and snow) is directly harmful to vegetation but also contributes to the fall in soil pH. Organic acids derived from the microbial breakdown of organic matter also lead to an increase in soil acidity.

The bacterial nitrification of **ammonia** to nitrate yields acid hydrogen ions. Consequently fertilizers containing ammonium salts prevent calcium from attaching to soil colloids and cause calcium loss in the **drainage water**. Other fertilizers have much less effect. Calcium and magnesium are plant nutrients and the soil's lime reserves are therefore gradually reduced by **crop removal**.

**Effects of liming**

When lime is added to an acid soil the calcium or magnesium replaces the exchangeable hydrogen on the soil colloid surface and neutralizes the soluble acids (*see* cation exchange). Eventually hydrogen ions are completely replaced by bases and **base saturation** is achieved, producing a soil pH of about 7. The influence of soil pH on plant **nutrient availability** is demonstrated in Figure 16.1. It can be seen that for mineral soils the most

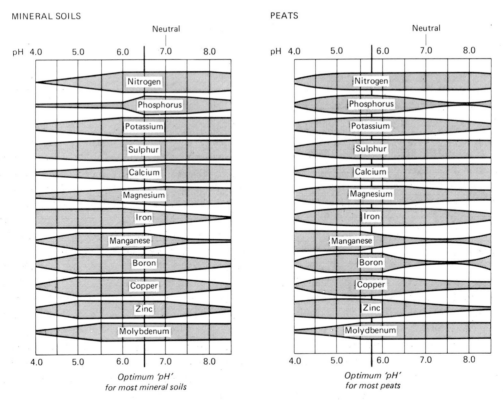

**Figure 16.1** *Effect of soil pH on nutrient availability. The availability of a given amount of nutrient is indicated by the width of the band. The growing media should be kept at a pH at which all essential nutrients are available. For most plants the optimum pH is 6.5 in mineral soils and 5.8 in peats.*

favourable level for ensuring the availability of all nutrients is pH 6.5, while peat soils should be limed to pH 5.8 in order to maximize overall nutrient availability. In acid conditions some nutrients such as manganese and other soil minerals such as aluminium may become toxic.

Beneficial **soil organisms** are affected by soil acidity and liming. A few soil-borne disease-causing organisms tend to occur more frequently on lime deficient soils (*see* clubroot), whereas others are more prevalent in well-limed soils. Calcium sometimes improves **soil structure** and **soil stability**. It is probable that this is mainly because it creates conditions favourable for decomposition of organic matter, yielding **humus**, and encourages **root activity**. Free lime in clay soils sometimes, but not always, leads to better crumb formation on drying and shrinking.

### Plant tolerance

Tolerance to soil pH and calcium levels varies considerably, but all plants are adversely affected when the soil becomes too acid. Table 16.1 shows the point below which the growth of common horticultural plants is significantly reduced. In the case of calcifuges (i.e. 'lime-hating' plants), the highest point before growth is affected by the presence of calcium should be noted: e.g. for *Rhododendrons* and some *Ericas* this is at pH 5.5.

### Lime requirement

The lime requirement of soil depends on the required rise in pH and the soil texture (*see* buffering capacity). **Lime requirement is expressed as the amount of calcium carbonate in tonnes per hectare required to raise the pH of the top 150 mm to the desired pH. A pH of 6.5 is recommended for temperate plants on mineral soils; pH 5.8 on peats.** The amount of a liming material needed to meet the lime requirement will depend on neutralizing value and sometimes fineness.

**Table 16.1**  Soil acidity and plant tolerance

| Plants | pH below which plant growth may be restricted on mineral soils |
|---|---|
| Beans | 6.0 |
| Cabbage | 5.4 |
| Carrots | 5.7 |
| Celery | 6.3 |
| Lettuce | 6.1 |
| Potato | 4.9 |
| Tomato | 5.1 |
| | |
| Apple | 5.0 |
| Blackcurrant | 6.0 |
| Raspberry | 5.5 |
| Strawberry | 5.1 |
| | |
| Carnation | 6.0 |
| Chrysanthemum | 5.7 |
| Daffodil | 6.1 |
| Hydrangea (pink) | 5.9 |
| Hydrangea (blue) | 4.1 |
| Rose | 5.6 |

### Liming materials

Liming materials can be compared by considering their ability to neutralize soil acidity, fineness, and cost to deliver and spread. **The neutralizing value (NV) of a lime indicates its potency. It is determined in the laboratory by comparing its ability to neutralize soil acidity with that of the standard, pure calcium oxide. A neutralizing value of 50 signifies that 100 kg of that material has the same effect on soil acidity as 50 kg of calcium oxide. The fineness** of the lime is important because it indicates the rate at which it affects the soil acidity (*see* surface area). It is expressed, where relevant, in terms of the percentage of the sample that will pass through a 100 mesh sieve. Liming materials commonly used in horticulture are listed below with some of their properties.

**Calcium carbonate** is the most common liming material. Natural soft chalk or limestone which is high in calcium carbonate is quarried and ground.

It is a cheap liming material, easy to store, and safe to handle. A sample in which 40 per cent will pass through a 100 mesh sieve can be used at the standard rate to meet the lime requirement. Coarser samples, although cheaper to produce, easier to spread and longer lasting in the soil, require heavier dressings. Shell sands have neutralizing values from 25 to 45. Whilst the purest samples can be used at nearly the same rate as calcium carbonate, twice as much of the poorer samples is required to have the same effect.

**Calcium oxide**, also known as quicklime, burnt lime, cob lime or caustic lime, is produced when chalk or limestone are very strongly heated in a lime kiln. Calcium oxide has a higher calcium content than calcium carbonate and consequently a higher neutralizing value. Pure calcium oxide is used as the standard to express neutralizing value (100) and the impure forms have lower values (usually 85–90). If used instead of ground limestone, only half the quantity should be applied. In contact with moisture, lumps of calcium oxide slake, i.e. react spontaneously with water to produce a fine white powder, calcium hydroxide, with release of considerable heat. Although rarely used now it should be used with care because it is a fire risk, burns flesh, and scorches plant tissue.

**Calcium hydroxide**, hydrated or slaked lime, is derived from calcium oxide by the addition of water. The fine white powder formed is popular in horticulture. It has a higher neutralizing value than calcium carbonate and its fineness ensures a rapid effect on the growing medium. Once exposed to the atmosphere it reacts with carbon dioxide to form calcium carbonate.

It should be noted that all forms of processed lime quickly revert to calcium carbonate when added to the soil. Calcium carbonate, which is insoluble in pure water, gradually dissolves in the weak carbonic acid of the soil solution.

**Magnesian limestone**, also known as Dolomitic limestone, is especially useful in the preparation of composts because it both neutralizes acidity and introduces magnesium as a nutrient. Magnesium limestone has a slightly higher neutralizing value (50–55) than calcium limestone, but tends to act more slowly.

### Lime application

Unless very coarse grades are used, lime raises the soil pH over a one to two year period, although the full effect may take as long as four years thereafter pH falls again. Consequently lime application should be planned in the planting programme. It is normally worked into the top 150 mm of soil. If deeper incorporation is required, the quantity used should be increased proportionally. The lime should be evenly spread and regular moderate dressings are preferable to large infrequent applications. Very large applications needed in land restoration work should be divided for application over several years.

Care should be taken that the surface layers of the soil do not become too acid even when the lower topsoil has sufficient lime. Top layers are the first to become depleted, with consequent effect on plant establishment.

Applications of organic manures or ammonium fertilizers should be delayed until lime has been incorporated. If mixed they react to release **ammonia** which can be wasteful.

### Decreasing soil pH

This is sometimes necessary for particular plants, e.g. *Ericas* requiring acid soils. In some circumstances it is appropriate to grow plants in a raised bed of acid peat or to work large quantities of peat into the topsoil. Either in conjunction with this approach or as an alternative, the base saturation of the mineral soil can be reduced by adding acids. Some acid industrial by-products can be used but the most usual method is to apply agricultural **sulphur** which is converted to sulphuric acid by soil micro-organisms. The sulphur requirement depends on the pH change required and the soil's buffering capacity. The application of large quantities of organic matter gradually makes soils more acid. Nitrogen fertilizers releasing **ammonia** considerably reduce soil pH over a period of years in outdoor soils (*see* soil acidity) and can be used in liquid feeding to offset the tendency of hard water to raise pH levels in composts.

## FERTILIZERS

Fertilizers are concentrated sources of plant nutrients which are added to growing media. **Straight fertilizers** are those which supply only one of the major nutrients: nitrogen, phosphorus, potassium or magnesium. The amount of nutrient in the fertilizer is expressed as a percentage. Nitrogen fertilizers are described in terms of percentage of the element nitrogen in the fertilizer, i.e. % N. Phosphate fertilizers have been described in terms of the equivalent amount of phosphoric oxide, i.e. % $P_2O_5$, or sometimes as percentage phosphorus, % P. Likewise potash fertilizers, i.e. % $K_2O$, or sometimes percentage potassium, % K. Magnesium fertilizers are described in terms of % Mg (*see* Table 16.2). The percentage figures clearly show the quantities of nutrient in each 100 kg of fertilizer.

**Compound fertilizers** are those which supply two or more of the nutrients nitrogen, phosphorus and potassium. The nutrient content expressed as for straight fertilizers is, by convention, written on the bag in the order nitrogen, phosphorus, potassium. For example, 20–10–20 denotes 20% N, 10% $P_2O_5$ and 10% $K_2O$. Fertilizer regulations require that further details of trace elements, pesticide content and phosphorus solubility should appear where applicable on the invoice.

Fertilizers and manures are available in many different forms. Generally the term **organic** implies that the fertilizer is derived from living organisms, whereas **inorganic** fertilizers are those

**Table 16.2** Nutrient analysis of fertilizers

| | N % | $P_2O_5$ (P) % | $K_2O$ (K) % | Mg % | Ca % | S % | Na % |
|---|---|---|---|---|---|---|---|
| Ammonium nitrate | 33–35 | | | | | | |
| Ammonium sulphate | 20–21 | | | | | 24 | |
| Bone meal | 3 | 20 (9) | | | | | |
| Calcium nitrate | 15.5 | | | | 20 | | |
| Calcium sulphate | | | | | 23 | 18 | |
| Chilean potassium nitrate | 15 | | 10 (8) | | | | 20 |
| Dried blood | 12–14 | | | | | | |
| Hoof and horn | 12–14 | | | | | | |
| Kieserite | | | | 16 | | | |
| Magnesium sulphate | | | | 10 | | 13 | |
| Meat and bone meal | 5–10 | 18 (8) | | | | | |
| Monoammonium phosphate | 12 | 37 (25) | | | | | |
| Phosphoric acid | | 54 (24) | | | | | |
| Potassium chloride | | | 59 (49) | | | | |
| Potassium nitrate | 14 | | 46 (38) | | | | |
| Potassium sulphate | | | 50 (42) | | | 17 | |
| Shoddy (wool waste) | 2–15 | 18–20 (8–9) | | | | | |
| Superphosphate | | 47 (20) | | | 20 | 12–14 | |
| Triple superphosphate | | | | | 14 | | |
| Urea | 46 | | | | | | |

derived from non-living material. However, in the context of organic growing it is necessary to look at specific requirements (Table 15.1).

### Application methods

Fertilizers are applied in several different ways. **Base dressings** are those which are incorporated in the growing medium. This can be achieved by **combine drilling** with seeds and fertilizer running into the same drill. In horticulture, however, **band placement** of fertilizers is far more common, involving equipment which drills the seeds in rows and places a band of fertilizer in parallel a few centimeters below and to one side. The risk of retarded germination or scorch of young plants due to high soluble fertilizers placed near seeds is thus avoided (*see* salt concentration). There is much less risk if fertilizer is **surface broadcast**, ie. scattered on prepared soil surface, or broadcast on the surface to be cultivated-in during the final stages of seed bed preparation. **Top-dressings** are fertilizers added to the soil surface but not incorporated. Such fertilizers must be soluble and not fixed by soil because the nutrient is carried to the roots by soil water. Nitrogen is the material most frequently applied by this method mainly because the large applications to crops require a base dressing and one or more top-dressings to minimize the risk of scorch and loss by leaching. Fertilizers are carefully incorporated in composts during the mixing of the ingredients (*see* compost mixing). **Liquid feeding** is the application of fertilizer diluted in water to the root zone. **Foliar feeding** is the application of a liquid fertilizer in suitably diluted form to be taken up through leaves. This technique is usually restricted to the application of trace elements.

### Formulations

**Quick acting** fertilizers contain nutrients in a form which plant roots can take up, and dissolve as soon as they come in contact with water, e.g.

ammonium nitrate and potassium chloride. Many are obtainable as powders or crystals. These are difficult to spread or place evenly but several are formulated this way to help in the preparation of liquid feeds. For this purpose they must be readily soluble and free of impurities which might lead to blockages in the feed lines. Some of the less soluble fertilizers, as well as lime, are spread in a finely divided form for maximum effect on soil. One of the major problems with many of the fertilizers is their hygroscopic nature, i.e. they pick up water from the atmosphere and create storage and distribution problems. Powdered forms in particular go sticky or 'cake' if dried out. Fertilizers formulated as **granules** are more satisfactory for accurate placement or broadcasting. They flow better, can be metered, and be thrown more accurately. **Prilled** fertilizers represent an improvement on granules because of their uniform spherical shape.

**Slow-release fertilizers** are those in which a large proportion of the nutrient is released slowly. Several of these fertilizers, such as rock phosphate, are insoluble or only slightly soluble and the nutrients are released only after many months, even years. Organic products are broken down by micro-organisms and the rate at which nutrients become available depends on their activity (*see* bacteria). Some slow-release artificial fertilizers, such as those based on urea formaldehyde, dissolve slowly in the soil solution whilst others are formulated in such a way that the soluble fertilizer they contain diffuses slowly through a resin coat, e.g. Osmocote, or sulphur coating, e.g. Gold-N. Some of these slow-release fertilizers have been formulated in such a way as to release nutrients at a rate which matches a plant's uptake and as such are sometimes referred to as **controlled-release** fertilizer. **Frits**, made from fine glass powders containing nutrient elements, are used either to release soluble materials slowly or to overcome the trace element problem caused by the narrow limits between deficiency and toxicity (*see* trace elements). Frits ease the difficulties experienced in mixing tiny amounts evenly through large volumes of compost. **Ion-exchange resins** release their nutrients by exchange with cations in the surrounding water. These resins help to overcome the problems of high salt concentration and leaching of nutrients from growing media based on inert materials (*see* aggregate culture).

Some plant nutrients are formed as **chelates** to maintain availability in extreme conditions where the mineral salt is 'locked up'. There are many different chelating or sequestrating agents selected for each element to be protected, and for each unavailability problem. Iron is chelated with EDDHA to form the product Chel 138 or Sequestrine 138 which releases the element in all soils including those with a high pH (*see* iron deficiency). EDTA, effective where there are high levels of copper, zinc and manganese, is used to chelate iron to be applied in foliar sprays.

## PLANT NUTRIENTS

### Nitrogen

Nitrogen is taken up as the nitrate and to a lesser extent the ammonium ion. Nitrate is converted to ammonium in the plant where it is utilized in the plant to form protein. Large quantities are used by plants and it is associated with vegetative growth. Consequently large dressings of nitrogen are given to leafy crops, whereas fruit, flower or root crops require limited nitrogen balanced by other nutrients to prevent undesirable characteristics occurring.

**Excess nitrogen** produces soft, lush growth making the plant vulnerable to pest attack and more likely to be damaged by cold. Very large quantities of nitrogen are undesirable since they harm the plant by producing high salt concentrations at the roots (*see* **conductivity**) and are lost by leaching. Large quantities are usually applied as a split dressing, e.g. some in base dressing and the rest in one or more top-dressings.

**Nitrates are mobile in the soil, which makes them vulnerable to leaching. In Britain it is assumed that all nitrates are removed by the winter rains so that virtually none is present until the soils warm up and nitrification begins or artificial nitrogen is applied** (*see* nitrogen cycle). Nitrates leached through the root zone may find their way into the groundwater which is the basis of the water supply in some areas. Nitrification also leads to the **loss of bases**; for every 1 kg N in the ammonia form that is oxidized to nitrate and leached, up to 7 kg of calcium carbonate or its equivalent is lost. Nitrogen is also lost from the root zone by denitrification, especially in warm, waterlogged soil conditions. When in contact with calcareous material, ammonium fertilizers are readily converted to ammonia gas which is lost to the soil unless it dissolves in surrounding water. For this reason urea or ammonia-based fertilizers should not be applied to such soils as a top dressing or used in contact with lime.

Nitrogen fertilizers used in horticulture and their nutrient content are given in Table 16.2. **Ammonium nitrate** is now commonly used in horticulture. In pure form it rapidly absorbs moisture to become wet; on drying it 'cakes' and can be a fire risk. Pure ammonium nitrate can be safely handled in polythene sacks and as prills. **Ammonium sulphate** has a very acid reaction in the growing medium.

**Urea** has a very high nitrogen content and in contact with water it quickly releases ammonia. Its use as a solid fertilizer is limited but it is utilized in liquid fertilizer or in foliar sprays. **Organic sources** of nitrogen, including dried blood, hoof and horn and shoddy, amongst others, are generally considered to provide slow-release nitrogen, but in warm greenhouse conditions decomposition is quite rapid.

### Phosphorus

Phosphorus is taken up by plants in the form of the phosphate anion $H_2PO_4^-$. Phosphorus is mobile in the plant and is constantly being recycled from the older parts to the newer growing areas. In practice this means that, although seeds have rich stores of phosphorus, phosphate is needed in the seedbed to help establishment. Older plants have a very low phosphate requirement compared with quick-growing plants harvested young.

**Most soils contain very large quantities of phosphorus but only a small proportion is available to plants.** The concentration of available phosphate ions in the soil water and on soil colloids is at its highest between pH 6 and 7. Phosphorus is released from soil organic matter by micro-organisms (*see* mineralization), but most of it and any other soluble phosphorus, including that from fertilizers, is quickly converted to insoluble forms by a process known as phosphate fixation. Insoluble aluminium, iron and manganese phosphates are formed at low pH and insoluble calcium phosphate at high pH. The carbonic acid in the vicinity of respiring roots (*see* respiration) and organisms in the rhizosphere facilitate phosphorus uptake.

The low solubility of phosphorus in the soil makes it virtually immobile, with the result that roots have to explore for it. Soils should be cultivated to allow roots to explore effectively; compacted or waterlogged areas deny plants phosphorus supplies. Phosphate added to the soil should be placed near developing roots (*see* band placement) in order to reduce phosphorus fixation and ensure that it is quickly found. If applied to the surface, phosphate fertilizers should be cultivated into the root zone. **Unlike soils, most artificial growing media have no reserves of phosphorus and when added in soluble form it remains mobile and subject to leaching.** Incorporating phosphorus in liquid feeds in hard water is complicated by the precipitation of insoluble calcium phosphates which lead to blocked nozzles. Slow-release phosphates are often selected in these situations to reduce losses and to eliminate the need for phosphorus in the liquid feeds.

Phosphorus nutrition used to be based on organic sources such as bones, but now phosphate fertilizers are mainly derived from rock phosphate ore. **Slow-acting forms** such as rock phosphate, bone meal and basic slag can be analysed in terms of their 'citric soluble' phosphate content, this being a good guide to their usefulness in the first season. Such materials should be finely ground to

enhance their effectiveness. These phosphates are applied mainly to grassland, tree plantings and in the preparation of herbaceous borders, to act as long-term reserves, particularly on phosphate deficient soils. Magnesium ammonium phosphate, calcium metaphosphate and potassium metaphosphate contain other nutrients but are valuable slow-release phosphates for use in soilless growing media. **Water-soluble phosphates** are produced by treating rock phosphates with acids. Superphosphate, derived from rock phosphate by treating with sulphuric acid, is composed of a water-soluble phosphate and calcium sulphate (gypsum), whereas triple superphosphate, derived from a phosphoric acid treatment, is a more concentrated source of phosphorus with less impurity. Both superphosphate and triple superphosphate are widely used in horticulture and are available in granular or powder form. Whilst they have a neutral effect on soil pH they tend to reduce the pH of composts. High grade monoammonium phosphate is used as a phosphorus source in liquid feeds because it is low in iron and aluminium impurities which lead to blockage in pipes and nozzles.

### Potassium

Potassium is taken up by the roots as the potassium cation and is distributed throughout the plant in inorganic form where it plays an important role in plant metabolism. For balanced growth the nitrogen to potassium ratio should be 1:1 for most crops but 2:3 for root and legumes. Leafy crops take up large amounts of potassium, especially when given large amounts of nitrogen. Where potassium supplies are abundant some plants, especially grasses, take up 'luxury' levels, i.e. more than needed for their growth requirements. Consequently, if large proportions of the plant are taken off the land, e.g. as grass clippings, there is a rapid depletion of potassium reserves.

**Potassium forms part of clay minerals and is released by chemical weathering. The potassium in soil organic matter is very rapidly recycled and**

exchangeable potassium cations held on the soil colloids and in the soil solutions are readily available to plant roots. Potassium and magnesium ions mutually interfere with uptake of each other. This ion antagonism is avoided when the correct ratio between 3:1 and 4:1 available potassium to magnesium is present in the growing medium. Availability of potassium is also reduced by the presence of calcium (*see* induced deficiency antagonism). Potassium is easily leached from sands low in organic matter and from most soilless growing media.

The main potassium fertilizers used in horticulture are detailed in Table 16.2. Although cheaper and widely used in agriculture, potassium chloride causes scorch in trees and can lead to salt concentration problems because the chloride ion accumulates as the potassium is taken up. Commercial potassium sulphate can be used in base dressings for composts but only the more expensive refined grades should be used in liquid feeding. More usually potassium nitrate is used to add both potassium and nitrate to liquid feeds, but is hygroscopic. Most potassium compounds are very soluble so that the range of slow-release formulations is limited to resin-coated compounds.

### Magnesium

Magnesium is an essential plant nutrient in leaves and roots and taken up as a cation. There are large reserves in most soils, especially clays, and those soils receiving large dressings of farmyard manure. Deficiencies are likely on intensively cropped sandy soils if little organic manure is used. Magnesium ion uptake is also interfered with by large quantities of potassium ions or calcium ions because of **ion antagonism**. Chalky and over-limed soils are less likely to yield adequate magnesium for plants.

Magnesium fertilizers include magnesian limestones containing a mixture of magnesium and calcium carbonate which raises soil pH (*see* liming). Magnesium sulphate as kieserite provides magnesium ions without affecting pH levels and in a pure form, Epsom salts, is used for liquid feeding and foliar sprays.

### Calcium

Calcium is an essential plant nutrient taken up by the plant as calcium cations.

Generally a satisfactory pH level of a growing medium indicates suitable calcium levels (*see* effects of liming). Gypsum (calcium sulphate) can be used where it is desirable to increase calcium levels in the soil without affecting soil pH. Deficiencies are infrequent and usually caused by lime being omitted from composts. Inadequate calcium in fruits is a more complex problem involving the distribution of calcium within the plant. Calcium nitrate or chloride solutions can be applied to apples to ensure adequate levels for safe storage (*see* plant tissue tests).

### Sulphur

Sulphur taken up as sulphate ions is a nutrient required in large quantities for satisfactory plant growth. It is not normally added specifically as a fertilizer because the soil reserves are replenished by re-circulated organic matter and air pollution. Furthermore, several fertilizers used to add other nutrients are in sulphate form, e.g. ammonium sulphate, superphosphate and potassium sulphate, and as such supply sulphur as well. As air pollution is reduced and fewer sulphate fertilizers are used, it is becoming necessary for growers to take positive steps to include sulphur in their fertilizer programme.

### TRACE ELEMENTS

Trace elements, also known as microelements or minor elements, are present in plants in very small quantities but are just as essential for healthy growth as major elements. Furthermore, they can be toxic to plants if too abundant. This means that rectifying deficiencies with soluble salts has to be undertaken carefully.

### Deficiencies

**Simple deficiencies** are those in which too little of the nutrient is present in the growing medium. Most soils have adequate reserves of trace elements and so simple deficiencies in them are uncommon, especially if replenished with bulky organic matter. Sandy soils tend to have low reserves and so too have several organic soils from which trace elements have been leached. In horticulture simple deficiencies of trace elements are mainly associated with growing in soilless composts which require careful supplementation.

**Induced deficiencies** are those in which sufficient nutrient is present but other factors such as soil pH or ion antagonism interfere with plant **nutrient availability**. On mineral soils boron, copper, zinc, iron and manganese become less available in alkaline soils, whereas molybdenum availability is reduced severely in soils with pH levels below 5.5 as shown in Figure 16.1. Trace element problems are aggravated in dry soils or where root activity is reduced by waterlogging, root pathogens or poor soil structure.

**Iron deficiency** is induced by the presence of large quantities of calcium and this **'Lime induced' chlorosis** (yellowing) occurs on overlimed soils and calcareous soils. The natural flora of chalk and limestone areas are calcicoles. Other plants grown in such conditions usually have a typically yellow appearance. Deficiencies can also be induced by high levels of copper, manganese, zinc and phosphorus. Top fruit and soft fruit are particularly susceptible as well as crops grown in complete nutrient solutions. The problem is overcome by using iron chelates.

**Boron deficiency** tends to occur when pH is above 6.8. It is readily leached from peat. Crops grown in peat are particularly susceptible when pH levels rise (Figure 16.1). Boron can be applied to soils before seed sowing in the form of borax or 'Solubor'.

**Manganese deficiency** is more frequent on organic and sandy soils of high pH. Plant uptake can be reduced by high potassium, iron, copper and zinc levels. Manganese availability is greatly increased at low pH and can reach toxic levels which most commonly occur after steam sterilization of acid, manganese-rich soils. High phosphorus levels can be used to reduce the uptake of manganese in these circumstances.

**Copper deficiency** usually occurs on peats and sands, notably reclaimed heathland, and in thin organic soils over chalk. High rates of nitrogen can accentuate the problem. Soils can be treated with copper sulphate or plants can be sprayed with copper oxychloride.

**Zinc deficiencies** are not common and are usually associated with high pH.

**Molybdenum deficiency** occurs on most soil types at a low pH. Availability becomes much reduced below pH 6 especially in the presence of high manganese levels. Cauliflowers are particularly susceptible and soils are limed to solve the problem. Sodium or ammonium molybdate can be added to growing media or liquid feeds where molybdenum supplies are inadequate.

## FERTILIZER PROGRAMME

The fertilizer applications required to produce the desired plant growth vary according to the type of plant, climate, season of growth and the nutrient status of the soil. General advice is available in many publications including those of the national advisory services and horticultural industries. Examples are given in Table 16.3.

### Growing medium analysis

The nutrient status of growing media varies greatly between the different materials and within the same materials as time passes. The nutrient levels change because they are being lost by plant uptake, leaching and fixation and gained by the weathering of clay, mineralization of organic matter, and the addition of lime and fertilizers.

There are many visual symptoms which indicate a deficiency of one or more essential nutrients but unfortunately by the time they appear the plant has probably already suffered a check in growth or change in the desired type of growth. The concentration of minerals in the plant and the nature of growth are linked so that **plant tissue analysis**, usually on selected leaves, can provide useful information, particularly in the diagnosis of some nutrient deficiencies; e.g. it is used to identify the calcium levels in apples in order to check their storage qualities. However, nutrient supply is usually assessed by analysis of the growing medium.

A **representative sample** of the growing medium is taken and its nutrient status determined. This involves extracting the **available nutrients** and measuring the quantities present. The **pH level** is determined and, where appropriate, the **lime requirement**. The **conductivity** of growing media from protected culture is also measured. The **nitrogen status** of soils is usually determined from previous cropping because, outdoors, nitrates are washed out over winter and their release from organic matter reserves is very variable. In protected planting, nitrate and ammonia levels are usually determined.

Results are often given in the form of an index number in order to simplify their presentation and interpretation. The ADAS soil analysis index is based on a ten point scale from 0 (indicating levels corresponding to probable failure of plants if nutrient is not supplied) to 9 (indicating excessively high levels of nutrient present).

**Fertilizer recommendations**

The results of the **growing medium analysis** are interpreted with the appropriate **nutrient requirement tables** to determine the actual amount of fertilizer to apply. These tables usually have growing medium nutrient status indices to aid interpretation and results are normally given in kg of nutrient per hectare or grams of nutrient per square metre (Table 16.3). In some cases the amount of named fertilizer required is stated, if another fertilizer is to be used to supply the nutrient the quantity needed must be calculated using the nutrient content figures (Table 16.2). It is important that throughout the fertilizer planning process the same units are used, i.e. % $P_2O_5$ *or* P%; $K_2O$ *or* % K. Conversion figures are:

$$\% \ P_2O_5 = \% \ P \times 2.29$$
$$\% \ P \quad = \% \ P_2O_5 \times 0.44$$
$$\% \ K_2O = \% \ K \times 1.20$$
$$\% \ K \quad = \% \ K_2O \times 0.83$$

**Sampling**

Normally only a small proportion of the whole growing medium is submitted for analysis and therefore it must be a **representative sample** of the whole. This is not easy because of the variability of growing media, particularly soils. It is recommended that each sample submitted for testing should be taken from an area no greater than 4 hectares. The material sampled must itself be uniform and so only areas with the same characteristics and past history should be put in the same sample. Irrespective of the area involved, from small plot to 4 hectare field, at least 25 subsamples should be taken by walking a zig-zag path avoiding the atypical areas such as headlands, wet spots, old paths, hedgelines, old manure heaps etc. (*see* Figure 16.2). The same amount of soil should be taken from each layer to a depth of 150 mm. This is most easily achieved with a soil auger or tubular corer.

Peat bags should be sampled with a cheese-type corer by taking a core at an angle through the planting hole on the opposite side of the plant to the drip nozzle from each of 30 bags chosen from an area up to a maximum of 0.5 hectares. Discard the top 20 mm of each core and if necessary take more than 30 cores to make up a one litre sample for analysis.

Samples should be submitted to analytical laboratories in clean containers capable of completely retaining the contents. They should be accompanied by name and address of supplier, the date of sampling, and any useful background

**Table 16.3**  Example of fertilizer requirements

**Carrots**

| N, P or K index | | 0 | 1 | 2 | 3 | 4 | over 4 |
|---|---|---|---|---|---|---|---|
| | | | | (kg/ha) | | | |
| Carrots, early bunching | N | 60 | 25 | Nil | – | – | – |
| | $P_2O_5$ | 400 | 300 | 250 | 150 | 125 | Nil |
| | $K_2O$ | 200 | 125 | 100 | Nil | Nil | Nil |
| Carrots, fen peats | N | Nil | Nil | Nil | – | – | – |
| (maincrop) other soils | N | 60 | 25 | Nil | – | – | – |
| all soils | $P_2O_5$ | 300 | 250 | 200 | 125 | 60 | Nil |
| | $K_2O$ | 200 | 125 | 100 | Nil | Nil | Nil |

Carrots on sandy soils respond to salt: 150 kg/ha Na (400 kg/ha salt) should be applied and potash reduced by 60 kg/ha $K_2O$. Salt must be worked deeply into the soil before drilling or be ploughed in.

**Dessert apples: mature trees**

| | Summer rainfall | Cultivated or overall herbicide | Grass/ herbicide strip | Grass | |
|---|---|---|---|---|---|
| | | (kg/ha per year) | | | |
| Nitrogen (N) | more than 350 mm | 30 | 40 | 90 | |
| | less than 350 mm | 40 | 60 | 120 | |
| P, K or Mg index | 0 | 1 | 2 | 3 | over 3 |
| | | (kg/ha per year) | | | |
| $P_2O_5$ | 80 | 40 | 20 | 20 | Nil |
| $K_2O$ | 220 | 150 | 80 | Nil | – |
| Mg | 60 | 40 | 30 | Nil | – |

For the first three years, fertilizer is not required by *young trees* grown in herbicide strips, provided that deficiencies of phosphate, potash and magnesium are corrected before planting by thorough incorporation of fertilizer.

information. All samples must be clearly identified. Further details of sampling methods in greenhouses or orchards, bags, pots, straw bales, water etc. are obtainable from the advisory services used. Remember, the result of the analysis can be no better than the extent to which the sample is representative of the whole.

## SOIL CONDUCTIVITY

The **soil solution** is normally a weaker solution than the plant cell contents. In these circumstances plants readily take up water through their roots by osmosis (*see* pages 11 and 12). As more salt, such as soluble fertilizer, is added to the soil

**Lettuce: base dressings for border soils under glass**

| Nitrate P, K or Mg index | Ammonium nitrate | Triple superphosphate | Sulphate of potash | Kieserite |
|---|---|---|---|---|
| | | g/m² | | |
| 0 | 30 | 150 | 160 | 110 |
| 1 | 15 | 140 | 110 | 80 |
| 2 | Nil | 130 | 50 | 30 |
| 3 | Nil | 110 | Nil | Nil |
| 4 | Nil | 80 | Nil | Nil |
| 5 | Nil | 45 | Nil | Nil |
| Over 5 | Nil | Nil | Nil | Nil |

Increase the nitrogen application by 50 per cent for summer grown crops. Lettuce is sensitive to low soil pH and the optimum pH is in the range 6.5 to 7.0. It is also sensitive to salinity, and growth may be retarded on mineral soils when the soil conductivity index is greater than 2.

solution, **salt concentrations** are increased and less water, on balance, is taken up by roots. When salt concentrations are balanced as much water passes out of the roots as into them. When salt concentrations are greater in the soil the roots are plasmolysed. The root hairs, then the roots, are 'scorched' i.e. irreversibly damaged, and the plant dries up.

Symptoms of high salt concentration above ground are related to the water stress created. Plants wilt more often and go brown at the leaf margin. Prolonged exposure to these conditions produces hard, brittle plants, often with a bluey tinge. Eventually severe cases become dessicated.

Salt concentration levels are measured indirectly using the fact that as salt concentration is increased the solution becomes a better conductor of electricity. The conductivity of soil solution is measured with a **conductivity meter**.

Salt concentration problems are most common where fertilizer salts accumulate, as in climates with no rainfall period to leach the soil and in protected culture. Periods when rainfall exceeds evaporation, as in the British Isles during winter, ensure that salts are washed out of the ground. Any plant can be damaged by applications of excess fertilizer and seedlings are very sensitive. Young roots can be scorched by the close proximity of fertilizer granules in the seedbed (*see* **band placement**).

In **protected culture** large quantities of fertilizer are used and residues can accumulate, particularly if application is not well adjusted to plant use. Sensitive plants such as lettuce are particularly at risk when following heavily-fed, more tolerant plants such as tomatoes or celery. Salt concentration levels should be carefully monitored and feeding adjusted accordingly, applying water alone if necessary. Soils can be **flooded** with water between plantings to leach excess salts. Large quantities of water are needed, but should be applied so that the soil surface is not damaged.

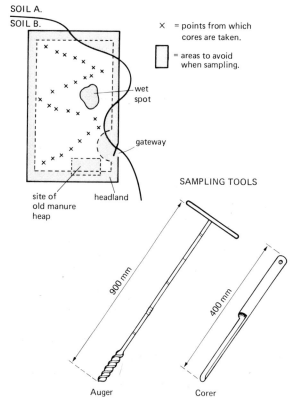

SAMPLING TOOLS

**Figure 16.2** *Sampling growing media.*
*Suitable tools to remove small quantities of growing media are corers or augers which have the advantage of removing equal quantities from the top and bottom of the sampled zone. The material to be sampled must be clearly identified, then 25 cores should be removed in a zig-zag that avoids anything abnormal.*

## PRACTICAL EXERCISES

1  Prepare a GCRI **general purpose potting compost** (*see* Table 17.1) with:
   (a)  all nutrients present.
   (b)  no nitrogen; exclude all nitrogen fertilizers and use potassium sulphate to supply potassium.
   (c)  no phosphorus; exclude all phosphate fertilizer.
   (d)  no potassium; exclude all potash fertilizer.
   (e)  no magnesium; exclude magnesian limestone and double the ground chalk level.
   (f)  no trace elements; exclude fritted trace elements.
   (g)  no lime; exclude chalk and replace magnesian limestone with one tenth by weight of kieserite or Epsom salts.
   (h)  different pH levels; using the mixture described in (g) and adding × 0.1, × 0.5, × 1, × 2 and × 10 the recommended ground chalk. Measure pH levels if possible.

   Label carefully and grow plants such as tomatoes in each of the above mixes. Note the growth in each and account for differences. Repeat with different plant species.

2  Compare different fertilizers based on the **cost per kg of the required nutrient**. Account for why the cheapest source of nutrient is not always selected.

## FURTHER READING

ADAS/ARC. *The Diagnosis of Mineral Disorders in Plants*, Vol. 1 *Introduction*, Vol. 2 *Vegetable crops*. General editor J. B. D. Robinson (HMSO, 1982), Vol. 3 *Glasshouse Crops*, G. Winsor and P. Adams (HMSO, 1987).

Archer, J. *Crop Nutrition and Fertilizer Uses*, 2nd edition (Farming Press, 1988).

Hay, R.K.M. *Chemistry for Agriculture and Ecology* (Blackwell Scientific Publications, 1981).

MAFF Reference Book No. 209 *Fertilizer Recommendations* (1988).

Roorda von Eysinga, J.P.N.L., and Smilde, K.W. *Nutritional Disorders in Chrysanthemums* (Centre for Agricultural Publishing and Documentation, Wageningen, 1980).

Roorda von Eysinga, J.P.N.L., and Smilde, K.W. *Nutritional Disorders in Glasshouse Tomatoes, Cucumbers and Lettuce* (Centre for Agricultural Publishing and Documentation, Wageningen, 1981).

Simpson, K. *Fertilizers and Manures* (Longman Handbooks in Agriculture, 1986).

# Alternatives to Growing in the Soil

*The soil has many advantages over alternative growing media, not least because it is usually the most available, and plants normally grow in it. Furthermore, it tends to be expensive to modify a soil or use an alternative growing medium. However, soil does have serious limitations for use in many aspects of horticulture and consequently the plants are frequently grown in suitable material put over the soil or, more commonly, in containers such as pots, liners, troughs, modules, cells etc.*

*In this chapter the principles underlying the use of containers and the mixes used in them are established, the nature of alternative materials and systems is examined and the advantages and disadvantages of the different approaches are discussed.*

## GROWING IN CONTAINERS

Whilst the importance of supplying water to plants in a restricted root volume is usually understood, the difficulties associated with achieving it whilst maintaining adequate **air-filled porosity (AFP)** are less well appreciated.

Roots require oxygen to maintain growth and activity. As temperatures rise, the plant requires more but the amount of oxygen that can be dissolved in water falls. Even in cool conditions, the oxygen that can be extracted from the water provides only a fraction of the root requirements. So, unless the plants have special modifications to transport oxygen through their tissues, as in aquatics, there has to be good gaseous movement through the growing medium. Many large interconnected pores allow rapid entry of oxygen.

It is generally agreed that 10–15% **air-filled porosity** is needed for a wide range of plants. Azaleas and epiphytic orchids require 20% or more, whereas others, including chrysanthemums, lilies and poinsettia, tolerate 5–10% AFP and carnations, conifers, geraniums, ivies and roses can be grown at levels as low as 2%.

Creating successful physical conditions depends on the use of components which provide a high proportion of macropores. Even in mixes which retain a good reservoir of water, very large quantities of water have to be applied over the course of a season so that the materials chosen must have very good **stability**; fine sand and silt soils collapse too quickly and reduce the size of the pore spaces. The sizes of the components used must be selected carefully to ensure that they create macropores, but also so that the gaps between the larger particles are not subsequently filled in by smaller particles. This is most easily achieved by using closely graded coarse particles. The reverse is achieved when combining many different-sized particles, as one would in mixing concrete, where the object is to minimize the air spaces as shown in Figure 17.1.

In addition to open ground or greenhouse borders, plants may be grown in pots, troughs, bags and other containers where restricted rooting makes more critical demands on the growing medium for air, water, and nutrients.

Ensuring that a growing medium in a container has adequate air-filled porosity is made difficult because water does not readily leave the container unless it is in good contact through its holes with similar-sized pore spaces. This is the case when placed on sand or capillary matting, but when stood out on gravel or wire the water will cling to the particles in the container (*see* water films). This can be tested by fully watering a pot of compost, holding it until it has finished dripping then touching the compost through a hole; normally a stream of water will run down the finger.

In fact the compost acts as a sponge and if at less than container capacity it will 'suck up' water from below. Furthermore, the lower layers remain almost saturated irrespective of the height or width of the container. This makes it particularly difficult to get good aeration in shallow trays, modules and blocks.

Providing the **nutrients** for the plant through a small volume means that, if supplied in soluble form in one application, the salt concentration produced is often too high, especially for seedlings (*see* conductivity). Consequently feeding has to be modified using split dressings, slow release fertilizers or liquid feeds.

## COMPOSTS

In  horticulture  the  growing  media  used  in

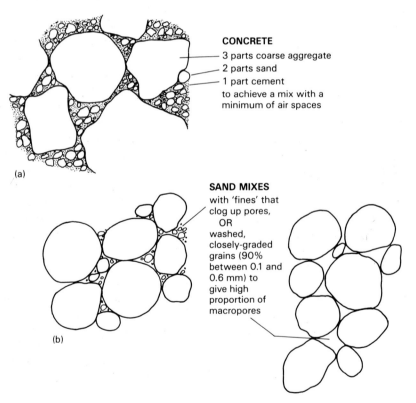

**CONCRETE**
- 3 parts coarse aggregate
- 2 parts sand
- 1 part cement

to achieve a mix with a minimum of air spaces

(a)

**SAND MIXES**
with 'fines' that clog up pores,
   OR
washed, closely-graded grains (90% between 0.1 and 0.6 mm) to give high proportion of macropores

(b)

**Figure 17.1** *Pore spaces in (a) concrete mix, (b) and (c) sand mixes.*

containers are usually referred to as 'composts'. These materials are also called plant substrates, plant growing media, or just 'mixes' or 'media'. Over the years growers have added a wide variety of materials such as leaf mould, pine needles, spent hops, old mortar, crushed bricks, composted animal and plant residues, peat, sand and grit to selected soils to produce a compost with suitable physical properties. To supplement the nutrient released from the materials in the compost, various slow-release organic manures or small dressings of powdered soluble inorganic fertilizers have been added to the mixtures to provide the necessary nutrition.

The correct physical and nutritional conditions are vital to successful growing in a restricted rooting volume. The most important development was as a result of the work done in the 1930s at the John Innes Institute. The range of composts that resulted from this work established the methods of achieving uniform production and reliable results with a single potting mixture suitable for a wide range of plant species.

## Loam composts

Loam composts, typified by John Innes composts, are based on loam sterilized to eliminate the soil-borne fungi (*see* damping off) and insects which largely caused the unreliable results from traditional composts. There is a risk of ammonia toxicity developing after sterilization of soil with pH greater than 6.5 or very high in organic matter (*see* nitrogen cycle). Induced nutrient deficiencies are possible in soils with a pH greater than 6.5 or less than 5.5. Furthermore, loam should have sufficient clay and organic matter present to give good structural stability. Peat and sand are added to further improve the physical conditions, the peat giving a high water holding capacity and the coarse sand ensuring free drainage and therefore good aeration. There are two main John Innes composts, one for seed sowing and cuttings, the other for potting.

**John Innes seed compost** consists of 2 parts loam, 1 part peat, and 1 part sand. Well-drained turfy clay loam with low nutrient and a pH between 5.8 and 6.5; undecomposed peat graded 3 mm to 10 mm with a pH between 3.5 and 5.0; and lime-free sand graded 1–3 mm should be used. 1200 g of superphosphate and 600 g of calcium carbonate are added to each cubic metre of compost. **John Innes potting (JIP) composts** consist of 7 parts by volume loam, 3 parts peat, and 2 parts sand. To allow for the changing nutritional requirements of a growing plant, the nutrient level is adjusted by adding appropriate quantities of JI base fertilizer which consists of 2 parts by volume hoof and horn, 2 parts superphosphate, and 1 part potassium sulphate. To prepare JIP 1, 3 kg JI Base and 600 g of calcium carbonate are added. To prepare JIP 2 and JIP 3, double and treble fertilizer levels respectively are used.

Whilst the standard JI composts are suitable for a wide range of species, some modification is required for some specialized plants. For example, calcifuge plants such as *Ericas* and *Rhododendrons* should be grown in a JI(S) mix in which sulphur is used instead of calcium carbonate.

All loam-based composts should be made up from components of known characteristics and according to the specification given. Such composts are well proven and are relatively easy to manage because of the water-absorbing and nutrient-retention properties of the clay present.

These composts are commonly used by amateurs, for valuable specimens, and for tall plants where pot stability is important; but loam-based composts have been superseded in horticulture generally by cheaper alternatives. The main disadvantage of loam-based composts has always been the difficulty in obtaining suitable quality loam as well as the high costs associated with steam sterilizing. Furthermore, the loam must be stored dry before use and the composts are heavy and difficult to handle in large quantities. Many loam-based composts currently produced have relatively low loam contents and consequently exhibit few of its advantages.

**Soilless or loamless composts**

These introduced the advantages of a uniform growing medium. Their component materials do not require steam sterilization, are lighter and cleaner to handle, and cheaper to prepare. Many have low nutrient levels which enables growers to manipulate plant growth more precisely through nutrition, but the control of nutrients is more critical as many components have a low **buffering capacity**.

**Peat** has until recently been the basis of most loamless composts. It is used alone or in combination with materials such as sand to produce the required rooting environment. Peats are derived from partially decomposed plants and their characteristics depend on the plant species and the conditions in which they are formed (*see* Chapter 15). Peats vary and respond differently to herbicides, growth regulants and lime. All peats have a high cation exchange capacity which gives them some buffering capacity. The less decomposed sphagnum peats have a desirable open structure for making composts and all peats have high water-holding capacities. Peat boards consist of highly-compressed peat which reduces transport costs. On wetting, the boards open up in the bags to form ideal rooting environments. Considerable efforts are being made to reduce the destruction of peatlands by finding alternatives to peat for use in horticulture. A list of some of the materials used are given in Table 17.1. Much progress has been made by using suitably processed bark or coconut fibre in composts. Along with several other organic sources they are waste-based and recycling them helps in conserving resources. All such alternatives must be free of toxins and pathogens, particularly those which may be a hazard to humans. In contrast, most of the inorganic alternatives are made from non-renewable resources (sand, loam) or consume energy in their manufacture (plastic foams, polystyrene) or both (vermiculite, perlite, rockwool). It seems unlikely that versatile peat will be replaced by a single alternative but rather different sectors will adopt substitutes best suited to their requirements.

**Table 17.1**   Alternatives to peat

| Organic materials | Inorganic materials |
| --- | --- |
| Pine | Expanded aggregates |
| Coir | Extracted minerals |
| Garden compost | Hydroponics |
| Heather/bracken | Perlite |
| Leafmould | Polystyrene |
| Lignite | Rockwool |
| Recycled landfill | Dredgings/warp |
| Refuse-driven humus | Vermiculite |
| Seaweed | Topsoil |
| Sewage sludge | |
| Spent hops and grains | |
| Spent mushroom compost | |
| Straw | |
| Vermicomposts | |
| Wood chips/fibre | |
| Woodwastes | |
| Wood fibre | |

**Sand and gravel** are used in composts, frequently in combination with peat. They have no effect on the nutrient properties of composts except by diluting other materials. They are used to change physical properties. As sand or gravel is added to lightweight materials the density of the compost can be increased, which is important for ballast when tall plants are grown in plastic pots. Sand is also used as an inert medium in aggregate culture.

It is important that the sands used should have low lime levels, otherwise they may induce a high pH and associated mineral deficiencies (*see* trace elements).

**Pulverized bark** has been used as a mulch and soil conditioner for many years. More recently it has been tried in compost mixtures as a replacement for peat. There are many different types of bark and they have different properties. Its problems include the presence of toxins, overcome by composting, and a tendency to 'lock-up' nitrogen (*see* Chapter 15) which can be offset by extra nitrogen in the feed. When composted with sewage sludge, a material suitable as a plant growing medium is produced. It is increasingly being incorporated into growing mixes in the attempt to reduce the use of peat.

**Coconut wastes** such as coir are proving to be useful in growing mixes. The material has good water-holding capacity, rewetting and air-filled porosity. It has buffering capacity because of a substantial cation exchange capacity. It has a pH between 5 and 6 which makes it suitable for a wide range of plants, but cannot replace peat directly in mixes for calcifuges. It has a carbon : nitrogen ratio of 80 : 1 which means that allowance has to be made for its tendency to lock up nitrogen.

**Vermiculite** is a mica-like mineral expanded to twenty times its original size by rapid conversion to steam of its water content. The finished product is available in several grades all of which produce growing media with good aeration and water holding properties. There is a tendency for the honeycomb structure to break down and go 'soggy'. Consequently, for long-term planting, it tends to be used in mixtures with the more stable peat or perlite. Some vermiculites are alkaline but the slightly acid samples are preferred in horticulture. Vermiculite has a high cation exchange capacity which makes it particularly useful for propagation mixes. Most samples contain some available potassium and magnesium.

**Perlite** is a mineral which is crushed and then expanded by heat to produce a white, lightweight aggregate The granules are non-porous but the rough surface holds more water than gravel or polystyrene balls. It tends to be used to improve aeration of growing media generally and the rewetting of peat. It is devoid of nutrients and has no cation exchange capacity. Graded samples may be used in aggregate culture (*see* page 190).

**Expanded polystyrene** balls or flakes provide a very lightweight inert material which can be added to soils or composts as a physical conditioner. It is non-porous and so reduces the water-holding capacity of the growing medium while increasing its aeration, thus making it less liable to water-logging when overwatered.

**Rockwool** is an insulation material derived from a granite-like rock crushed, melted, and spun into threads. The resulting slabs of lightweight spongey, absorbent, inert and sterile rockwool provide ideal rooting conditions with high water-holding capacity and good aeration. Shredded rockwool can be used in compost mixes. Its pH is high but easily reduced by watering with a slightly acid nutrient solution. It is frequently used in tomato and cucumber production and film-wrapped cubes are available for plant raising and pot plants. It is necessary to use a complete nutrient feed (*see* aggregate culture). It has some buffering capacity but this is very low on a volume basis. The main problem areas lie in calcium and phosphorus supply and the control of pH and salt concentration. Some rockwool has been formulated with clay to overcome some of these problems. This increases its cation exchange properties, making it very suitable for interior landscaping.

Rockwool is also available in water-absorbent and water-repelling forms. Mixtures of these enables formulators to achieve the right balance between air-filled porosity, water holding and capillary lift. Rockwool is available in slabs but also as granules which provides a flexible alternative for those who produce their own mixes.

**Plastic foams** of several different types are becoming popular for propagation because of their open porous structure. They are available as flakes and balls for addition to composts or as cubes into which the cuttings can be pushed.

### Compost formulations

Materials alone or in combination are prepared and mixed to achieve a rooting environment which is free from pests and disease organisms and has adequate air-filled porosity, easily available water, and suitable bulk density for the plant to be grown.

While lightweight mixes are usually advantageous, 'heavier' composts are sometimes formulated to give pot stability for taller specimens. This should not be achieved by compressing the lightweight compost but by incorporating denser materials such as sand. Quick-growing plants are normally the aim and this is obtained by loosely filling containers with the correct compost formulation and settling it with applications of water. Firming with a rammer reduces the total pore space whilst increasing the amount of compost and nutrients in the container. The reduction in available water and increase in soluble salt concentration leads to slower growing, harder plants (*see* conductivity).

The addition of nutrients has to take into account not only the plant requirements but also the nutrient characteristics of the ingredients used. Most soilless composts require trace element supplements and many, including those based on peat, need the addition of all major nutrients and lime.

Glasshouse Crops Research Institute Composts nos. 1, 2 and 3 are **general purpose potting composts** based on a peat/sand mix (*see* Table 17.2).

They contain different combinations of nutrients and consequently their storage life differs. Compost no. 3 has a slow-release phosphate, removing the need for this element in a liquid feed (*see* phosphorus).

The **GCRI seed compost** contains equal parts by volume of sphagnum peat and fine, lime-free sand. To each cubic metre of seed compost is added 0.75 kg of superphosphate, 0.4 kg potassium nitrate and 3.0 kg calcium carbonate.

**Compost mixing**

It is most important when making up the desired compost formulation to achieve a uniform product and, commercially, it must be undertaken with a minimum labour input. The ingredients of the compost must be as near as possible to the specification for the chosen formulation. Materials must not be too moist when mixing because it then

**Table 17.2**   GCRI composts

| Constituents | Seed compost | | Potting composts | | |
| | | | *Urea formaldehyde types*[1] | | *high P type*[2] |
| | | general use | Winter use | Summer use | |
| | | | **1** | **2** | **3** |
| **Peat : sand** | | | | | |
| (% by volume) | 50 : 50 | 75 : 25 | 75 : 25 | 75 : 25 | 75 : 25 |
| **Base dressings** | | | | | |
| (kg/m$^3$) | | | | | |
| Ammonium nitrate | Nil | 0.4 | Nil | Nil | 0.2 |
| Urea formaldehyde (a) | Nil | Nil | 0.5 | 1.0 | Nil |
| Magnesium ammonium phosphate | Nil | Nil | Nil | Nil | 1.5 |
| Potassium nitrate | 0.4 | 0.75 | 0.75 | 0.75 | 0.4 |
| Normal superphosphate | 0.75 | 1.5 | 1.5 | 1.5 | Nil |
| Sulphate of potash | Nil | Nil | Nil | Nil | Nil |
| Ground chalk | 3.0 | 2.25 | 2.25 | 2.25 | 2.25 |
| Ground magnesian limestone | Nil | 2.25 | 2.25 | 2.25 | 2.25 |
| Fritted trace elements (WM225) | Nil | 0.4 | 0.4 | 0.4 | 0.4 |

[1] Composts containing urea formaldehyde should not be stored longer than seven days.

[2] For longer term crops where there is a risk of phosphorus deficiency and liquid feeding with phosphate is not desired, use commercial magnesium ammonium phosphate. This also contains 11 per cent $K_2O$.

becomes impossible to achieve an even distribution of nutrients.

There are several designs of **compost mixer**. Continuous mixers are usually employed by specialist compost mixing firms and require careful supervision to ensure a satisfactory product. Batch mixers of the 'concrete mixer' design are produced for a wide range of capacities to cover most nursery needs. Many of the bigger mixers have attachments which aid filling. Emptying equipment is often linked to automatic pot-filling machines.

Ingredients used in loamless composts do not normally require partial sterilization, but **sterilizing equipment** is certainly needed to prepare loams for inclusion in loam-based composts. Where steam is used it is injected through perforated pipes on a base plate and rises through the material being sterilized. In contrast a steam–air mixture injected from the top under an air-proof covering is forced downwards to escape through a permeable base.

Storage of prepared composts should be avoided if possible and should not exceed three weeks if slow-release fertilizers are incorporated. If nitrogen sources in the compost are mineralized, **ammonium ions** are produced followed by a steady increase in **nitrates** (*see* Chapter 15). These changes lead to a rise in compost pH followed by a fall. As nitrates increase, the salt concentration rises towards harmful levels (*see* conductivity). Peat-based composts can become infested during storage by sciarid flies.

## PLANT CONTAINERS

The characteristics of the container affect the root environment as does the standing-out area. Clay pots are porous and water is lost from the walls by evaporation. Consequently clay pots dry out quicker than plastic ones, especially in the winter. This can help improve **air-filled porosity** although air does not enter through the walls. The higher evaporation rate also keeps the clay pots slightly cooler, which can be beneficial in hot conditions.

Likewise the contents of white plastic pots can be as much as 4°C lower than in other colours. Pots of white or light green plastic can transmit sufficient light to adversely affect root growth and encourage algal growth.

Paper and compressed peat containers have become popular because they can be planted directly. Some materials decompose more rapidly than others and there can be a temporary 'lock-up' of nitrogen but most peat containers are now manufactured with added available nitrogen. It is essential that the surrounding soil is kept moist after planting or the roots fail to escape from the dry wall.

### Blocks

Blocks are made of a suitable compressed growing medium into which the seed is sown with no container or simply a net of polypropylene. Aeration tends to be poorer than in pots but this is a successful means of growing some vegetables on a large scale.

### Modules

Increasingly traditional seedbed bare-rooted or block transplant techniques have become replaced by raising a wide variety of plants in modules. A module is made by adding a loose growing medium mix to a tray of cells. The cells are variously wedge or pyramid-shaped, so designed to enable a highly mechanized transplanting process to be used. Fine, free-flowing mixes of peat, polystyrene or bark are used to fill the cells, which have large drainage holes and no rim to hold free water. Roots in the wedge-shaped cells are 'air-pruned', which encourages secondary root development. 'Plugs' are mini-modules in which each transplant develops in less than $10\,cm^3$ of growing medium and are used for bedding plants as well as tomato production. The rate of establishment is largely determined by the water stress experienced by the transplant. Irrigation of the module or plug is found to be more successful than applying water to the surrounding growing medium.

## Hydroponics

Hydroponics (water culture) involves the growing of plants in water. The term often includes the growing of plants in solid rooting medium watered with a complete nutrient solution which is more accurately called '**aggregate culture**'. Plants can be grown in nutrient solutions with no solid material so long as the roots receive oxygen and suitable anchorage and support is provided (*see* Chapter 13).

The advantages of hydroponics compared with soil in temperate areas include accurate control of the nutrition of the plant and hence better growth and yield. There is a constant supply of available water and oxygen to the roots. Loss of water and nutrients through drainage is minimal in recirculating systems and evaporation is greatly reduced. There can be a reduction in labour and growing medium costs and a quicker 'turn round' time between crops in protected culture. The disadvantages include the costs of construction and the controls of the more elaborate automated systems.

## Aggregate culture

This relies on a complete nutrient solution to supply the plant. The solution is broken up into water films by a solid medium such as sand or gravel. Several types of expanded clay used in the building industry such as Leca have been used. Smooth but porous granules 4–8 mm in diameter, giving a capillary rise of about 100 mm, are used to create an ideal rooting environment with a dry surface which makes it an attractive method of displaying house plants (*see* Figure 17.2). All forms of aggregate culture require feeding with all essential minerals. Trace element deficiencies occur less frequently when clay aggregates are used.

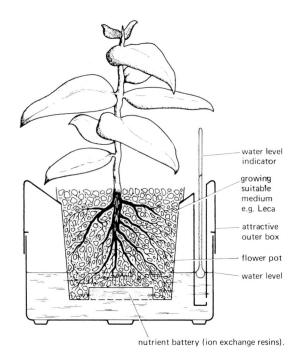

water level indicator

growing suitable medium e.g. Leca

attractive outer box

flower pot

water level

nutrient battery (ion exchange resins).

**Figure 17.2** *Plant pots with water reserves.*
*Plants grown in a variety of growing media can be fitted with reservoirs which supply water by capillarity. A water level indicator is frequently incorporated and in some systems the nutrients are supplied from ion exchange resins. While this system can be used for any pot size it is particularly attractive in large displays.*

Ion exchange resins are an ideal fertilizer formulation in these circumstances because the nutrients are released slowly, remove harmful chlorides and fluorides from irrigation water, and aid pH control.

## Nutrient film technique (NFT)

This is a method of growing plants in a shallow stream of nutrient solution continuously circulated along plastic troughs or gullies. The method is commercially possible because of the development of relatively cheap non-phytotoxic plastics to form the troughs, pipes and tanks (*see* Figure 17.3). There is no solid rooting medium and a mat of roots develops in the nutrient solution and in the

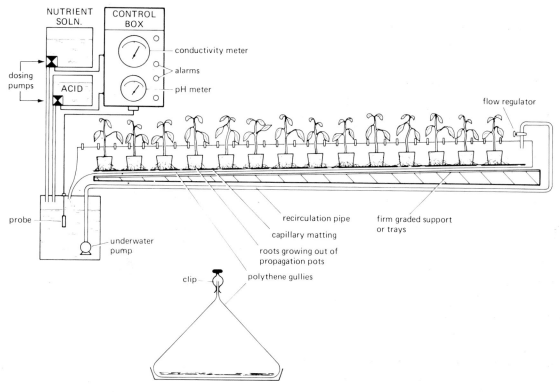

**Figure 17.3** *Nutrient film technique layout*
*The nutrient solution is pumped up to the top of the gullies. The solution passes down the gullies in a thin film and is returned to the catchment tank. The pH and nutrient levels in the catchment tank are monitored and adjusted as appropriate.*

moist atmosphere above it. Nutrient solution is lifted by a pump to feed the gullies directly or via a header tank. The ideal flow rate through the gullies appears to be 4 litres per minute. The gullies have a flat bottom, often lined with capillary matting to ensure a thin film throughout the trough. They are commonly made of disposable black/white polythene set on a graded soil or on adjustable trays. There must be an even slope with a minimum gradient of 1 in 100; areas of deeper liquid stagnate and adversely affect root growth (*see* aerobic conditions).

The nutrient solution can be prepared on site from basic ingredients or proprietary mixes. It is essential that allowance is made for the local water quality, particularly with regard to the microelements such as boron or zinc which can become concentrated to toxic levels in the circulating solution. The nutrient level is monitored with a conductivity meter and by careful observation of the plants. Maintenance of pH between 6 and 6.5 is also very important. Nutrient and pH control is achieved using, as appropriate, a nutrient mix, nitric acid or phosphoric acid to lower pH and, where water supplies are acid, potassium hydroxide to raise pH. Great care and safety precautions are necessary when handling the concentrated acids during preparation.

The commercial NFT installations have automatic control equipment in which conductivity and pH meters are linked to dosage pumps. The high and low level points also trigger visual or audible alarms in case of dosage pump failure. Dependence on the equipment may necessitate the grower installing failsafe devices, a second lift pump, and a standby generator. Despite the dominance of automatic control equipment and reduced labour by this method, 'plantsmanship' still remains a vital factor for success.

## SPORT SURFACES

The specifications for sports playing surfaces are such that turf has increasingly given way to artificial alternatives typified by the trend toward playing 'lawn' tennis on 'clay' courts. This is partly attributable to maintenance requirements, but at the higher levels of sport it is because the users or the management expect play to continue with a minimum of interference by rainfall. The usual problem is that the soil in which the turf grows does not retain its structure under the pounding it receives from players and machinery, especially when it is in the wet plastic state. Turf is still

preferred by many, but to achieve the high standards required it has to be grown in a much modified soil (*see also* sand slitting) or, increasingly, in an alternative such as sand.

The most extreme approach is to grow the turf in pure sand isolated from the soil, sometimes within a plastic membrane. The high cost of these methods is such that it is only used to create small areas such as golf greens.

Normally the existing topsoil is removed from the site and the subsoil is compacted to form a firm base and graded to carry water away to drains. Drainage pipes are laid, above which is placed goes a drainage layer usually consisting of washed, pea-sized gravel, as shown in Figure 17.4. Because it is considerably coarser than the sand placed on it, this layer prevents the downward percolation of water (*see* water films) and creates a perched watertable. This helps to give the root zone a large reserve of available water whilst ensuring that gravitational water, following heavy rain or excess irrigation, is removed very rapidly.

A 25–30 cm root zone of free-draining sand is

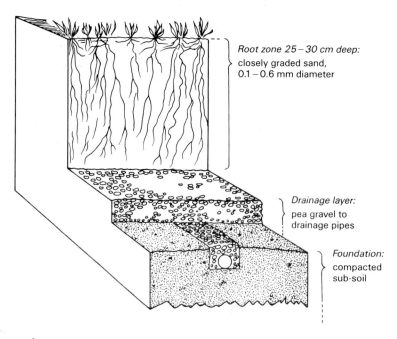

*Root zone 25–30 cm deep:*
closely graded sand,
0.1–0.6 mm diameter

*Drainage layer:*
pea gravel to
drainage pipes

*Foundation:*
compacted
sub-soil

**Figure 17.4** *Pure sand root zone.*

placed uniformly over the drainage layer, evenly consolidated. Allowance has to be made for continued settling over the first year. It is essential that the sand used has a suitable particle size distribution, ideally 80–95% of the particles being between 0.1 and 0.6 mm diameter. A minimum of 'fines' is essential to avoid the clogging up of the pores in the root zone. Sometimes a small amount of organic matter is worked into the top 5 cm to help establish the grass, although success is probably as easily achieved with no more than regular light irrigation and liquid feeding.

Some very sophisticated all-sand systems, such as the Cellsystem, are constructed so that the root zone is subdivided into bays with vertical plastic plates and supplied with drains that can be closed so that the water in each of them can be controlled. Tensiometers are used to activate valves that allow water back into the drainage pipes to sub-irrigate the turf.

## PRACTICAL EXERCISES

1   Prepare a GCRI **general purpose potting compost** with the recommended nutrients added to the recommended peat/sand mix but to other combinations also:
    (a)   75% by volume peat : 25% by volume sand.
    (b)   50% by volume peat : 50% by volume sand.
    (c)   25% by volume peat : 75% by volume sand.
    (d)   100% peat .
    (e)   100% sand.
    (f)   a sandy loam soil.
    Label carefully and grow plants such as tomatoes in each of the above mixes. Note the growth in each and account for differences. Repeat with different plant species.
2   Grow a selection of plant species in a range of compost ingredients and different mixes.
3   Grow a selection of plant species in a GCRI **general purpose potting compost** firmed to different extent.

## FURTHER READING

Bunt, A.C. *Media and Mixes for Container Grown Plants* (Unwin Hyman, 1988).
Cooper, A. *The ABC of NFT* (Grower Books, 1979).
Handreck, K.A., and Black, N.D. *Growing Media for Ornamentals and Turf* (New South Wales University Press, 1984).
Pryce, S. *The Peat Alternatives Manual* (Friends of the Earth, 1991).
Smith, D.L. *Rockwool in Horticulture* (Grower Books, 1987).

# Index